Edward Jimmy Pandia Yañez
Luis Alberto Melendez Ruiz
Nancy Ramos Maquera

Natural Soil Stabilization with Addition of Pneumatic Tire Rubber

Edward Jimmy Pandia Yañez
Luis Alberto Melendez Ruiz
Nancy Ramos Maquera

Natural Soil Stabilization with Addition of Pneumatic Tire Rubber

Stabilization of Natural Soil with Addition of Granular Rubber from Tires Madre de Dios

ScienciaScripts

Imprint
Any brand names and product names mentioned in this book are subject to trademark, brand or patent protection and are trademarks or registered trademarks of their respective holders. The use of brand names, product names, common names, trade names, product descriptions etc. even without a particular marking in this work is in no way to be construed to mean that such names may be regarded as unrestricted in respect of trademark and brand protection legislation and could thus be used by anyone.

Cover image: www.ingimage.com

This book is a translation from the original published under ISBN 978-620-2-16848-9.

Publisher:
Sciencia Scripts
is a trademark of
Dodo Books Indian Ocean Ltd. and OmniScriptum S.R.L publishing group

120 High Road, East Finchley, London, N2 9ED, United Kingdom
Str. Armeneasca 28/1, office 1, Chisinau MD-2012, Republic of Moldova, Europe
Printed at: see last page
ISBN: 978-620-7-69719-9

Copyright © Edward Jimmy Pandia Yañez, Luis Alberto Melendez Ruiz, Nancy Ramos Maquera
Copyright © 2024 Dodo Books Indian Ocean Ltd. and OmniScriptum S.R.L publishing group

AFFIDAVIT OF AUTHORSHIP

By means of this document, I, EDWARD JIMMY PANDIA YAÑEZ , identified with National Identity Card N° 41946689 with address at Jr. Parde de Miguel 977, in the district of Tambopata, province of Tambopata, department of Madre de Dios, former student of Alas Peruanas University declare under oath that:

I am the author of the research entitled: **STABILIZATION OF THE NATURAL SOIL SUBGRADE WITH ADDITION OF GRANULAR RUBBER OF TIRES IN THE JR. TUPAC AMARU, TAMBOPATA DISTRICT, MADRE DE DIOS, 2022,** which I present on the fifteenth day of November 2023, before this institution for the purpose of obtaining the academic degree of Civil Engineer.

This research has not been previously presented or published by any other researcher or by the undersigned for any other academic degree or professional title. I declare that all ideas, text, figures, formulas, tables or others that correspond to the undersigned or to others have been duly cited in full respect of copyright. I declare that I know and submit to the current legal and regulatory framework related to such responsibility.

I declare under oath that the data and information presented belong to the reality studied, that they have not been falsified, adulterated, duplicated or copied. That I have not committed scientific fraud, plagiarism or authorship vices; otherwise, I exempt Alas Peruanas University from any responsibility and I declare myself as the only responsible.

EDWARD JIMMY PANDIA YAÑEZ

GENERAL

TITLE

STABILIZATION OF THE NATURAL SOIL SUBGRADE WITH THE ADDITION OF GRANULAR RUBBER FROM TIRES IN JR. TUPAC AMARU, TAMBOPATA DISTRICT, MADRE DE DIOS, 2022.

AUTHOR

EDWARD JIMMY PANDIA YAÑEZ

Academic-Professional School of Civil Engineering.

Faculty of Engineering and Architecture.

U.A.P. Madre de Dios Branch.

TYPE OF RESEARCH

According to type: Applied research

According to Level: Explanatory

According to design: Experimental

LINE OF RESEARCH

Soil Mechanics, Geotechnics and Pavements

LOCATION

Tambopata district, Tambopata province, Madre de Dios department

THANK YOU

I express my gratitude to all the people and professionals who gave me their support at each stage of the research process. I acknowledge the fundamental support of my mother in my academic development, and in a special way, I thank my advisor and the UAP University.

INDEX

SUMMARY

The research addresses the application of tire granular rubber as an innovative technique to improve subgrade properties in natural soils. The objective of the research was to determine how the use of tire granular rubber influences subgrade stabilization. The study focuses on analyzing how this addition affects the physical and mechanical characteristics of the soil, examines in detail the stabilization process, evaluating the strength, compressibility and behavior of the soil modified with granular rubber. Specific tests and trials are carried out to quantify and compare the properties of the natural soil and the stabilized soil, providing key data for decision-making in infrastructure design and construction.

In addition, environmental aspects related to tire recycling are explored, considering the possibility of using products derived from these wastes to improve soil conditions. The research provides valuable insights into sustainability and environmental impact.

The research approach adopted was of an applied nature with an explanatory level, using an experimental design. The sample consisted of three excavations in Jr. Tupac Amaru. The technique used to collect data was experimental observation, and the instrument used was an observation guide that included soil analysis forms. The results show that the addition of granular rubber in the stabilization of the subgrade in Jr. Tupac Amaru, Tambopata District, Madre de Dios, is highly significant. Practical recommendations are provided based on the results obtained, offering guidelines for the successful implementation of this strategy in real projects.

Key words: Granular rubber, consistency limits, modified Proctor, and CBR.

ABSTRACT

The research addresses the application of granular tire rubber as an innovative technique that allows improving the properties of the subgrade in natural soils. The objective of the investigation was to determine how the use of granular tire rubber influences the stabilization of the subgrade. The study focuses on analyzing how this addition affects the physical and mechanical characteristics of the soil, examining in detail the stabilization process, evaluating the resistance, compressibility and behavior of the soil modified with granular rubber. Specific tests and trials are carried out to quantify and compare the properties of natural soil and stabilized soil, providing key data for decision making in infrastructure design and construction.

In addition, environmental aspects related to tire recycling are explored, considering the possibility of using products derived from these wastes to improve soil conditions. The research provides valuable perspectives on sustainability and environmental impact.

The research approach adopted was of an applied nature with an explanatory level, using an experimental design. The exhibition consisted of three excavations at Jr. Tupac Amaru. The technique used to collect data was experimental observation, and the instrument used was an observation guide that included soil analysis formats. The results show that the addition of granular rubber in the stabilization of the subgrade in Jr. Tupac Amaru, Tambopata District, Madre de Dios, is highly significant. Practical recommendations are provided based on the results obtained, offering guidelines for the successful implementation of this strategy in real projects.

Keywords: Granular rubber, consistency limits, modified Proctor, and CBR.

INTRODUCTION

Improving the properties of natural soil is a topic of growing interest in civil engineering and road and pavement construction. Soil subgrade stabilization is essential to ensure the durability and safety of road infrastructure.

An innovative perspective in this area involves the inclusion of recycled and environmentally friendly materials, such as granular rubber derived from recycled tires. This practice not only addresses the problem of waste tire disposal, but also has great potential to improve soil characteristics. The introduction of granular rubber into natural soil can help reduce expansion, increase bearing capacity and deformation resistance, while promoting environmental sustainability.

The focus of this research is to explore the feasibility and advantages of stabilizing the subgrade of natural soil by adding granular rubber from recycled tires. Through experimental methods and detailed analysis, we seek to evaluate the impact of this addition on the mechanical and geotechnical properties of the soil, with the purpose of determining its capacity to support loads and provide a solid base for pavement construction. In addition, aspects related to sustainability and the environment will be analyzed, contributing to an integral decision making process in the design of road infrastructure.

The purpose of this study is to deepen the topic of subgrade stabilization of natural soil through the incorporation of granular rubber from recycled tires, presenting it as an effective and sustainable solution in the field of road engineering.

The Madre de Dios region is characterized by clayey soil, rich in fine material and with high plasticity. This type of soil becomes extremely unstable in the presence of moisture, which poses considerable challenges for infrastructure construction, particularly in the case of roads. The subgrade, being composed of this type of soil, requires improvement or replacement to ensure the stability and performance of the road.

In wet conditions, these soils tend to behave in a soft manner and, during the drying process, undergo volume changes that result in settlement in various areas of the road. In order to address this problem, this research study incorporated granular tire rubber in proportions of 5%, 10% and 15% into the clayey soils. Laboratory analyses and evaluations were then conducted to study the impact of this mixture. The main objective of this work is to provide clayey soils with a more stable and durable consistency without deviating from current regulations and standards.

The results obtained in this study are expected to contribute to the development of more efficient and environmentally friendly practices in road and pavement construction and maintenance.

The objective of this research is to determine to what extent the use of granular rubber from tires influences the stabilization of the natural soil subgrade in Tupac Amaru Jr., Tambopata District, Madre de Dios, during the year 2022.

CHAPTER I: PROBLEMATIC REALITY

1.1. Description of the Problem Reality

The city of Madre de Dios relies primarily on mining, chestnut harvesting, the timber industry, and agriculture. As a result, during the harvest season, the local roads experience a considerable and constant flow of trucks transporting these products. Due to the lack of adequate quality control in the selection of materials and the execution of these roads in the city, they suffer accelerated deterioration, resulting from the combination of rainy weather conditions and intense traffic activity.(Pinto, 2018)

Clay soils are noted for their remarkable plasticity, which implies that they can undergo substantial transformations as they alternate between wet and dry states. Due to the presence of minute particles with a remarkable water-holding capacity, they tend to expand when saturated and contract when dehydrated. This variation in the expansion and contraction of these soils poses a significant risk to infrastructure, such as roads, as it can trigger deformations in the surface and cause damage to the integrity of the road. These effects can manifest themselves in the form of potholes and cracks, which in turn result in uncomfortable and unsafe traffic for drivers. As a consequence, there are increased costs associated with road maintenance and repair.

Prior to the construction process, soil amendments, such as chemical stabilization or controlled compaction, can be used to mitigate the expansion and contraction inherent in clay soils. Another suggested alternative involves the incorporation of granulated rubber from the shredding of recycled tires into the natural soil. This addition of recycled tire rubber is mixed with the soil in order to enhance its properties, including bearing capacity, wear resistance and stability.

According to Esteve (2012), it has been established that the way in which tires are produced and the obvious challenges related to their storage and disposal, once they have been used, currently constitute one of the most pressing environmental problems of recent decades worldwide. It is recognized that the manufacture of a tire requires considerable amounts of energy, equivalent to half a barrel of crude oil for the production of a truck tire. Furthermore, if these tires are not properly recycled, they generate significant environmental pollution when they end up as part of uncontrolled landfills.

Although the practice of tire recycling can be environmentally positive by reducing the accumulation of tire waste, it raises concerns regarding the potential release of chemicals and the durability of shredded rubber when incorporated into the soil. The addition of granular rubber often involves costs, so it is essential to evaluate the benefits associated with improved soil conditions and durability of infrastructure to determine the economic viability of this technique.

The incorporation of granular rubber as a stabilization measure has the potential to increase the bearing capacity and wear resistance of the soil, which is particularly important in areas where soils present significant challenges. Proper subgrade stabilization can contribute to increased infrastructure longevity, resulting in decreased costs associated with long-term maintenance.

The investigation will be carried out at Jr. Tupac Amaru:
- Department: Madre De Dios
- Province: Tambopata
- District: Tambopata
- Natural Region: Jungle

Photo 1: *Project location*

1.2. Problem Formulation

1.2.1. General Problem

P.G. To what extent does the use of granular tire rubber influence the stabilization of the natural soil subgrade at Jr. Tupac Amaru, Tambopata District, Madre de Dios , 2022?

1.2.2. Specific Problems

PE1: How the addition of granular rubber from tires improves rá the consistency limits for the stabilization of the subgrade in Jr. Tupac Amaru, Tambopata District, Madre de Dios, 2022 ?

PE2: In what way the addition of tire granular rubber will improve the modified Proctor for the stabilization of the subgrade in Jr. Tupac Amaru, Tambopata District, Madre de Dios, 2022?

PE.3. To what extent the addition of granular rubber from tires will improve the CBR in the stabilization of the subgrade in Jr. Tupac Amaru, Tambopata District, Madre de Dios, 2022?

1.3. Project Objectives

1.3.1. General Objective

O.G. Determine how the use of granular rubber from tire tires influences the stabilization of the natural soil subgrade in Jr. Tupac Amaru, Tambopata District, Madre de Dios, 2022 .

1.3.2. Specific objectives

SO.1 Determine how the addition of tire granular rubber will improve the Consistency Limits for subgrade stabilization on Jr. Tupac Amaru, Tambopata District, Madre de Dios, 2022 .

OE.2 Determine how the addition of tire granular rubber will improve the modified Proctor for subgrade stabilization in Jr. Tupac Amaru, Tambopata District, Madre de Dios, 2022 .

SO.3 Determine how the addition of tire granular rubber will improve CBR in subgrade stabilization on Jr. Tupac Amaru, Tambopata District, Madre de Dios, 2022 .

1.4 Rationale and importance of the investigation

1.4.1 Justification

The natural ground beneath, known as subgrade, often does not have the necessary qualities to ensure the stability and bearing capacity required for construction and infrastructure projects. The theoretical basis for this justification focuses on the need to enhance the characteristics of the soil, including its bearing capacity and wear resistance, in order to ensure the durability of structures and the safety of roadways.

Soil stabilization represents a field of civil engineering supported by solid theoretical foundations. The inclusion of granular rubber conforms to these principles and is underpinned by an understanding of how certain materials can enhance soil properties.

The theoretical basis for this approach is based on the recognition that used tires pose an environmental challenge and the imperative need to discover sustainable applications for their recycling. The incorporation of granular rubber from recycled tires into the subgrade is based on the premise of beneficially reusing an unused material, helping to reduce the accumulation of tires in landfills. This theory is based on the consideration of sustainability and the reduction of adverse effects on the environment.

From a theoretical perspective, subgrade stabilization theoretically leads to decreased long-term maintenance costs by strengthening the durability of structures. This theory supports the idea that the initial investment in stabilization can translate into substantial savings in terms of future expenditures for repairs and ongoing care.

The incorporation of granular rubber from recycled tires into the subgrade has the potential to significantly enhance the bearing capacity of the soil. This concrete improvement means that roads and structures built on this subgrade can withstand heavier loads and constant traffic, resulting in improved performance and durability.

In practical situations, stabilization using granular rubber can reduce the propensity of the subgrade to deform over time, especially in variable weather conditions. This is essential to preserve the integrity of road and structure surfaces, which in turn reduces repair and maintenance costs over time. Subgrade stabilization with granular rubber can lead to long-term cost savings. The initial investment in subgrade improvement translates into less need for repairs and maintenance, thus generating financial savings.

From a practical perspective, the reuse of recycled tires as stabilization material effectively addresses the environmental concern related to the accumulation of tires in landfills. This is not only beneficial from an ecological perspective, but can also be a cost-effective solution for waste management.

The methodology should include a field and laboratory study approach to evaluate the behavior of the natural soil as well as the effectiveness of granular rubber stabilization. Specific tests will be required, such as laboratory analysis of soil samples and in situ tests at the construction site.

Key parameters to be evaluated and measured as part of the methodology should be identified and selected. This could include bearing capacity, wear resistance, soil deformation and other relevant factors.

The methodology should allow the comparison of results between the subgrade stabilized with granular rubber and the subgrade without stabilization. This implies the collection of data before and after the application of the technique to evaluate its impact.

1.4.2 Importance

Subgrade stabilization with granular rubber can reduce the need to replace or excavate large amounts of natural soil, which saves natural resources and reduces the costs of constructing and maintaining road infrastructure.

The use of recycled materials in civil engineering applications is an area of growing interest and offers opportunities for innovation in road design and construction. Research in this field opens up new perspectives for road engineering and pavement construction.

The reuse of discarded tires as granular rubber to stabilize the subgrade of natural soil is a sustainable practice that addresses the problem of tire waste management and contributes to the reduction of waste in landfills, as well as to the prevention of environmental pollution.
The addition of granular rubber significantly improves the mechanical properties of the soil, such as its load-bearing capacity, resistance to deformation and bearing capacity.

This has a direct impact on the quality and durability of road infrastructure, which in turn contributes to safer and more economical operation of roads and pavements.

1.4.3 Feasibility

The focus on environmental sustainability has become one of the most significant trends in the field of civil engineering and road construction. The use of granular rubber derived from recycled tires not only solves the problem of waste management, but also has the potential to decrease the carbon footprint and reduce the environmental impact compared to conventional methods of soil stabilization. Subgrade stabilization with granular rubber could have a positive impact on the economics of road construction by reducing the need for expensive imported materials and extensive excavation.

A subgrade that benefits from the incorporation of granular rubber can extend the life and durability of road infrastructure, which in turn reduces the need for long-term repairs and maintenance. This leads to a lower impact on the environment and promotes a more sustainable infrastructure.

1.5 Research limitations

The costs of laboratory analyses represent a significant constraint, as access to university laboratories is not available to carry out such analyses independently.

CHAPTER II: THEORETICAL FRAMEWORK

2.1. Background of the investigation

2.1.1 International Level

Patiño, J. (2017) conducted the research "*Soil stabilization using recycled rubber additions*". Thesis to opt for the degree of Ph.D. in

Civil Engineering, Faculty of Civil Engineering, Universidad Católica de Santiago de Guayaquil, Ecuador. In this thesis work, the stabilization process of two different types of soil using recycled rubber is carried out. Tests will be carried out with the same type of rubber, which comes from recycled tires; the only difference between the two types of rubber used lies in the amount of recycling processes to which they have been subjected. In addition, will carry out sorting processes for the two types of soil that will be used in the specimen tests. Tests will be carried out using two types of specimens: pure soil and soil mixed with rubber. These tests will be carried out in order to determine the resistance through the CBR (California Bearing Ratio) test and the density through the modified Proctor test, following the standards established by ASTM, which will be the fundamental protocols in this research. The results obtained in the laboratory will provide the necessary information to evaluate the effectiveness of soil stabilization through the soil-rubber mixture, considering the different types of mixtures that will be carried out .

Laica, J. (2016) conducted the research entitled "*Influence of the inclusion of recycled polymer (rubber) on the mechanical properties of a sub* base" . Thesis to opt for the degree of Civil Engineer, Faculty of Civil and Mechanical Engineering, Engineering Career, Technical University of Ambato, Ecuador. The report, created as part of an experimental study, has as its main purpose to improve the mechanical characteristics of a class 3 subbase by incorporating recycled polymer, in this case, rubber. To carry out this research, we began by collecting the necessary materials, which include the subbase and the recycled rubber polymer. In a first step, the physical-mechanical properties of the subbase were evaluated to determine if the material meets the requirements established by AASHTO and ASTM standards. The subbase was obtained from Constructora Alvarado Ortiz, located at the intersection of Arq.

Lecorbusier and Socrates Streets. The recycled rubber polymer was also obtained from Proneumacosa, located at Km 12 of the Panamericana Norte in the "La Avelina" sector. Compaction, modified Proctor and California Bearing Ratio (CBR) tests were carried out with the incorporation of rubber in different proportions. Finally, the results obtained from both the sample in its natural state and the sample with the addition of rubber in different percentages were compared. The results showed that as the amount of rubber in our subbase increases, the resistance decreases significantly .

Álvarez, S. (2020) conducted the research "*Use of pulverized rubber granules from used tires as a solution to reinforce soft subgrade soils in the Sabana de Bogotá*". Faculty of Civil and Environmental Engineering, Civil Engineering, Universidad Antonio Nariño, Bogotá, Colombia. Where the addition of pulverized rubber from disused tires was analyzed as an efficient solution to reinforce the soft subgrade soils found in the Bogotá Savanna. It reaches the conclusions that through various investigations carried out in different parts of the world on the use of recycled tires in soils that lack the capacity to support road structures, it has been evidenced that a recycled tire contains several materials that can be used to strengthen a subgrade soil with stability problems. Rubber powder, in particular, is presented as an option to reinforce unstable soils, since it can improve some of their mechanical properties, such as shear strength, cohesion and friction angle, especially in clayey soils. In addition, it has been observed that mixing 4% rubber powder with the unsuitable soil results in a significant increase in the California Bearing Ratio (CBR). In order to apply this methodology to the soils of the city of Bogotá, the properties and behavior of the materials present in the deposits of this area have been examined and characterized in detail. This has made it possible to identify the areas of the city where it would be feasible to use rubber powder to strengthen the soil. Comparison between traditional methods of soft soil improvement and the approach proposed in this report has revealed economic and environmental advantages for the city. However, due to the scarcity of research in Colombia on this particular topic, it is not possible to state with complete certainty whether rubber powder is suitable for use in the capital's soils. Nevertheless, this raises a new line of research in the country that could boost future studies related to this topic .

Ravichandran, PT, Prasad, AS, Krishnan, KD & Rajkumar, PRK (2016). wrote the article *"Effect of addition of shredded rubber from scrap tires on stabilization of weak soils"*. International Journal of Engineering, Department of Civil Engineering, SRM University, Kattankulathur - 603203, Tamil Nadu, India. The purpose of this research was to evaluate the feasibility of using granulated rubber powder as an additive to improve the strength of soils with low bearing capacity. Two types of problematic clayey soils were stabilized by adding different percentages of granulated rubber (5%, 10%, 15% and 20%). The strength properties of the stabilized soils were evaluated, focusing on the increase of the percentage of granulated rubber up to 10%, which was analyzed by California Bearing Ratio (CBR) tests. In addition to the improvement in strength, the effects of this type of stabilizer and variations in drainage capacity were examined. The incorporation of granular rubber in both soil types resulted in favorable changes in permeability. With 10% granulated rubber, an increase of 161% in the CBR value was observed for soil A1 and 130% for soil A2. These results indicate that both strength modification and permeability improvement contribute to more effective stabilization of clay soils. The increase in the CBR value of the stabilized soil has the potential to substantially reduce the overall pavement thickness, which in turn can result in a significant decrease in the overall costs associated with road construction.

2.1.2 National Level

Rojas, R. (2019) in his research entitled *"Subgrade improvement incorporating recycled granular rubber in Bonavista Avenue, Carabayllo, Lima - 2019 "*. Thesis to opt for the professional degree of Civil Engineer in the Faculty of Engineering, Academic Professional School of Civil Engineering, Universidad César Vallejo. Where it indicates that, over many years, road works have experienced numerous failures that decrease their useful life, resulting in additional maintenance costs. These problems are due to several factors, among which is the inadequate construction and resistance of the subgrade. On the other hand, the growth of solid waste, particularly scrap tires, and their inadequate management have generated environmental pollution in our surroundings. In this context, the objective of this research is to determine how the incorporation of recycled granular rubber from tires influences soil properties. The study was carried out in a section of Bonavista Avenue, located in the district of Carabayllo. This area is at subgrade level and presents a soil in poor condition. The research approach was based on the scientific method, using a quantitative approach that involved data collection to address the research questions. It is classified as applied research, since theories were applied for the benefit of the study area, with an explanatory level and an experimental design, where one of the variables was deliberately manipulated. The study population was the existing soil on Bonavista Avenue, and samples were taken from two 1.5 meter deep pits. The results of the improvement of the subgrade properties were not favorable, since the incorporation of recycled granular rubber failed to improve the compaction, strength and expansion of the analyzed soil. In view of the above, it is recommended to consider the use of recycled granular rubber in applications other than soil mechanics, in order to reduce environmental contamination in more suitable areas .

Cueto (2018) in his research work "*Propuesta técnica para estabilizar talud con neumáticos reciclados, trocha carrozable Hualituna - Curva Gervasio - región Junín*". Thesis for the degree of Civil Engineer, Universidad Peruana Los Andes. The objective was to determine a technical proposal for the stabilization of a slope using recycled tires in the Hualituna - Curva Gervasio dirt road in the Junin Region. The research was of an applied nature, since theoretical knowledge was applied to address existing problems, although it did not have an experimental design. The study population consisted of the embankments along the carriageway, and the sample ranged from kilometers 01+570 to 01+575. The results of this research concluded that the proposed technique using recycled tires would allow the stabilization of the slope on the Hualituna - Curva Gervasio dirt road. In addition, the use of recycled tires with a diameter of 0.72 m and an inner diameter of 0.45 m, filled with sterile material composed of clayey gravel type soil was proposed.

Huamán, R. & Muguerza, K. (2019) in their research "*Influence of granulated rubber in cohesive soils related to the penetration resistance property (CBR), 2019*". Thesis to opt for the Professional Degree of Civil Engineering, Faculty of Engineering, Academic Professional School of Civil Engineering, Universidad César Vallejo. In this research project, it was examined how granulated rubber affects the penetration resistance in cohesive soils of Huayllay - Huaychao. The approach of this thesis is considered applied due to its objective of addressing practical problems. The research focuses on describing the influence of granulated rubber and uses a quasi-experimental cross-sectional research design. The study area covered the Huayllay - Huaychao section in the province of Pasco, and soil samples were collected from three soil pits. These samples were taken to the laboratory, where it was studied how different proportions of granulated rubber (5%, 10% and 15%) affected the penetration resistance when added to the dry weight of the soil. In total, 12 CBR tests were performed to evaluate the penetration resistance and determine the optimum proportion of granulated rubber. The results obtained for test pit 3 indicate a SUCS classification of low plasticity clay (LC). An increase in the CBR was observed with the 5% and 10% rubber percentages, but not with 15%. The 10% granular rubber was identified as the optimum ratio, as it increased the soil strength from 5.2% to 12.2%. On the other hand, pits 1 and 2 were classified as gravel with clay (GC) according to SUCS, and a steady decrease in CBR was observed as each percentage of rubber was added, being from 26.27% to 20.16% and from 34.06% to 28.5%, respectively. In conclusion, it can be stated that granulated rubber improves cohesive soils, as long as the parameters established by the Peruvian standard MTC Suelos y Geotecnia - 2013 are met.

Casimiro, V. & Melgarejo, K. (2022) in their research "*Cohesive soil stabilized with granulated rubber to improve the physical-mechanical properties of a subgrade in rural areas*". Thesis for the degree of Civil Engineer, Faculty of Engineering, Professional School of Civil Engineering, Ricardo Palma University. In this study, the use of granulated rubber as a stabilizing agent in cohesive soils was investigated, evaluating its physical-mechanical properties, such as plasticity, density, moisture and bearing capacity. The methodology employed was deductive and was based on the review of more than 30 research papers related to the use of this by-product of discarded tires as a stabilizing agent in cohesive soils. This review provided a significant amount of information on the use of granulated rubber for the purpose of preventing future problems in road infrastructure projects. In addition, it examined how granulated rubber affects each of the properties of the cohesive soil and how these properties relate to each other. To carry out this research, soil samples were collected in the district of San Luis de Shuaro, province of Chanchamayo, Junín, which presented similar characteristics to those obtained in the review of data from various authors. The objective of this study was to analyze the behavior of cohesive soil when granulated rubber is added as a stabilizing agent. The experimental results indicated that the Shrinkage Limit increases with the incorporation of granulated rubber. The Maximum Dry Density reaches its maximum value when the soil is in its natural state, decreasing as granulated rubber is added. In relation to the CBR in critical conditions (soaked state), a decrease in resistance was observed after the incorporation of granulated rubber. This does not optimize the physical-mechanical properties of the soil, since it does not meet the minimum criteria required to be considered a good subgrade according to the Highway Manual (MC-05-14) .

2.2. Theoretical basis

2.2.1. Soils

Soil is a mixture of elements including minerals, organic matter, gases, liquids and various organisms, all of which contribute to sustaining life on our planet. Soil undergoes constant evolution, driven by a series of physical, chemical and biological processes, including weathering caused by erosion. The main task of stabilization is concentrated in areas with soft soil consistency, such as loam, clay, peat or organic soils, in order to achieve suitable engineering properties. According to Sherwood's observations, granular materials with small particles tend to be more susceptible to stabilization because of their large surface area compared to their particle size. On the other hand, clayey soils are noted for their large surface area due to the flattened and elongated shape of their particles. However, silty soils can be sensitive to minor changes in moisture, which can lead to complications during the stabilization process (Sherwood, 1993).

Within civil engineering, soil plays a fundamental role in all projects, since it acts as the base on which all loads and stresses involved will be distributed. However, not all soils are suitable for this purpose because their geotechnical properties may not be appropriate, leading to the need to modify them to increase their strength and durability (Delgado, C. & Mormontoy, V., 2021).

2.2.2. Clay soils

Clay soil is characterized by the predominant presence of clay particles compared to particles of other sizes. Clay consists of very small mineral particles, less than 0.001 mm in diameter, as opposed to larger particles, such as silt and sand, which vary in size from smaller to larger. In clay soils, there is a combination of silt and sand, but the proportion of clay is predominant and may vary according to the specific soil characteristics (Quesada, 2008).

2.2.2.1. Variables influencing the expansive behavior of clayey soils

Several factors influence the expansive behavior of soils.

Changes in Moisture Content. Variations in moisture in the active zone of the soil profile are crucial in determining whether the soil will swell or shrink.

Initial Moisture Condition. An expansive soil that has been previously dewatered shows a higher affinity for water or high suction compared to the same soil with a higher initial moisture content. In other words, the lower the initial moisture, the greater the tendency to expand.

Initial Stress Conditions. A considerable reduction in the initial stresses of a soil stratum can cause significant relaxation, resulting in more marked volumetric changes.

Initial Dry Density. When soil density is high, it generally implies that soil particles are closer together, which increases the repulsive forces between them and thus the propensity for swelling when the soil absorbs water.

Drainage and Other Water Sources. The presence of broken pipes or soil irrigation can change the moisture content of the soil.

Soil Particle Structure and Organization. Clays with a flocculated structure tend to show a greater tendency to swell compared to those with a dispersed structure.

Stress History. Soils that have previously experienced significant stresses (overconsolidated soils) tend to be more expansive than those that have been subjected to normal stresses.

Influence of Climate. The balance between evaporation and precipitation in a given area can have a significant impact on soil moisture.

Mineralogy of clays. Clay minerals have varied expansive characteristics, and the expansion capacity of the soil is related to the type and amount of clay minerals present. Clay minerals belonging to the Smectite group (such as Montmorilonite) and Vermiculite are responsible for significant volumetric changes. Although Illites and Kaolinites are rarely expansive, they can experience volume changes when their particle size is very small (less than 0.1 micrometers).

Plasticity. Generally speaking, soils that exhibit plastic behavior over a wide range of moisture contents and have a high Liquid Limit have a greater potential for shrinkage and swelling.

Groundwater chemistry. Cations such as sodium, calcium, magnesium and potassium dissolved in water adhere to the surface of clays as exchangeable cations to balance surface electrical charges. The type of exchangeable cation influences the expansive properties of the soil.

Suction in Soil. Suction in soils manifests itself through negative pore pressure in unsaturated soils. The greater the suction, the greater the degree of swelling.

Water Table and Groundwater Conditions. Fluctuations in the water table can contribute to changes in soil moisture content.

Effect of Vegetation. Trees, shrubs and grasses absorb moisture from the soil, creating zones of differential soil moisture.

Soil Profile. The thickness and location of an expansive stratum in the soil profile determine the magnitude and rate of the swelling process.

Soil Permeability. The presence of cracks and fissures in the soil, which increases its permeability, can accelerate water migration and thus the rate of soil expansion.

Temperature An increase in temperature causes moisture to move to cooler areas, which can influence the expansive behavior of the soil under pavements or buildings.

2.2.1.2. Properties of Clay Soils

Specific surface area. Refers to the calculation of the area encompassing both the external surface and, if present, the internal surface of the constituent particles. In clays, a remarkably high specific surface area is found, which plays a fundamental role in the interaction between the solid particles and the surrounding liquid.

Plasticity. The main characteristic of clay soils is their plasticity, which originates mainly due to the shape and size of the particles. The relationship between the amount of water and the amount of clay is essential, since water acts as a lubricating agent between the clay sheets, allowing them to slide in response to stresses applied by an external load. Plasticity is typically measured by determining Atterberg limits, which will be detailed later.

Hydration and expansion. The expansion phenomenon originates as water is absorbed into the space separating the clay sheets. This process can be explained as follows: when water penetrates and causes a greater separation between the sheets, electrostatic repulsion forces are generated between the sheets, which contributes to the expansion process and can lead to a complete separation between some of them.

Thixotropy. Thixotropy refers to the property of certain clays that, when kneaded, acquire a liquid consistency; however, when resting, they recover their cohesion and behave again as a solid. This thixotropic behavior is manifested when the clay contains a specific level of moisture close to its liquid limit. Conversely, when the moisture reaches a level close to the plastic limit, the clay will not exhibit its thixotropic characteristic.

Absorption capacity. Absorption capacity is closely related to soil texture, including specific surface area and porosity. Here, two distinct physical processes can be distinguished: the absorption process, which mainly involves physical phenomena, such as capillary retention; and the adsorption process, which involves a chemical interaction between the clay and the absorbed liquid. Due to its high porosity, clay exhibits a significant adsorption capacity.

Cation exchange capacity. This phenomenon is reversible and refers to the ability to exchange the ions fixed on the external surface of the crystals, the interlayer spaces and other internal areas of the structures for ions present in the surrounding aqueous solutions. This characteristic is crucial, since the mechanical properties of clays vary depending on the amount of cations present in their adsorption complexes. Different types of bound cations are associated with different thicknesses of adsorbed film, which in turn affect the plasticity and strength properties of the soil. This process occurs without substantially altering the structure of the solid. Therefore, controlled cation exchange is used to mechanically improve soils.

2.2.3. Subgrade

Subgrade refers to the layer of soil beneath the base of roads, pavements or other infrastructure structures. Its main role is to provide support and firmness to the base layer and pavement, thus transmitting the loads and stresses generated by traffic to the underlying soil. To ensure the strength and stability necessary to maintain the integrity of the structure to be built on it, the subgrade is selected and prepared. It can be composed of various types of soils and materials, and its quality plays an essential role in the durability and safety of the infrastructure.

2.2.3.1. Characterization of subgrades

To carry out the characterization of the materials used in the subgrade, it is necessary to perform excavations, specifically pits that are at least 1.5 meters deep. These pits should be located alternately and in a longitudinal direction, maintaining an approximately uniform distance between them.

The zone at the bottom of the subgrade, at a depth of no less than 0.60 meters, must be made up of soils capable of supporting the required loads, i.e., with a CBR (California Bearing Ratio) equal to or greater than 6%. On the other hand, if these strata do not comply with this range and have a CBR of less than 6% (which is considered a low quality or inadequate subgrade), it will be necessary to improve their resistance. At this stage, the engineer will carry out a detailed analysis and offer possible improvement solutions according to the guidelines established by the MTC (Ministry of Transport and Communications) in 2014.

2.2.4. Granular rubber

Rubber is a substance derived from hydrocarbons found in tree latex. By adding acetic acid or exposing it to high temperatures, the substance is solidified and separated from other components in minimal quantities. The result is raw rubber, which has a sticky and viscous texture, showing resistance and fragility at low temperatures, and flexibility at high temperatures. It is important to note that once the rubber has been subjected to stretching processes, it does not recover its original properties.

In 1839, Charles Goodyear discovered that by mixing rubber with sulfur and heating it to 100°C, the sulfur chemically combines with the rubber, generating significant improvements. This results in a rubber that does not crack when cold, does not deform with heat and does not become sticky. In addition, if stretched or manipulated, the rubber has the ability to recover its original properties.

The different types of synthetic rubber have a common origin in molecules such as butadiene, isoprene or their derivatives, all with a fundamentally similar structure (Castro, 2008, p. 45).

Granulated rubber exhibits characteristics of longevity, resistance to aging and easy maintenance.

As regards its physical characteristics, granulated rubber comes in the form of small black particles with a diameter of approximately 4 mm. It is important to note that its production does not have a negative impact on the environment and it has moderate elasticity, a non-slip surface, good water drainage capacity and good resistance to abrasion.

From a mechanical perspective, granulated rubber shows remarkable shear strength according to the results of triaxial tests. In addition, it has the ability to absorb vibrations and is flexible.

In terms of particle size, granulated rubber can be considered as a form of sand, as its particles are in the range of 0.075 mm to 4.75 mm in diameter.

The stabilization of the poor subgrade using granulated rubber reduces the thickness of the engineered pavement and prolongs the service life of the pavement. This technique involves the modification of the subgrade soil, taking advantage of the properties of granulated rubber, which is lightweight and shows high shear strength. In addition, it contributes to solving the problem of improper tire disposal and minimizes environmental pollution (Juliana et al., 2020, p. 2).

2.2.4.1. Rubber classification

Butadiene - Styrene. This material is composed of 75% butadiene and 25% styrene, and its production is carried out by a free radical process. It finds application in the manufacture of tires for vehicles.

Isobutylene Rubber - Isoprene. This material is similar to natural rubber in its performance, but does not have the same flexibility. It is characterized for being highly resistant to oxidation and corrosive products, besides having a low permeability to gases. For this reason, it is used in tire inner tubes and is known to be difficult to vulcanize.

Natural Rubber. It originates from various plants that generate a milky liquid, known as latex, when subjected to a cut in the trunk, and its color is generally white.

Synthetic Rubber. This material, similar to natural rubber, is produced artificially by chemical processes, specifically polymerization. After its manufacture, synthetic rubber undergoes the vulcanization process.

Neoprene. Neoprene is a type of synthetic rubber developed in the early stages of Carothers' research. It is noted for its high resistance to heat and chemicals such as oil and petroleum. It is commonly used in applications such as pipelines to transport oil and as insulation for cables.

Polybutadiene. Polybutadiene is a synthetic rubber widely used in the manufacture of tires due to its high wear resistance. It is formed through the polymerization process of the corresponding monomer.

2.2.4.2. Rubber properties

Shore hardness. Refers to the elastic response of rubber when subjected to impact against a hard surface. This measure evaluates the elastic recovery capacity of materials.

Elasticity. Refers to the ability of a material to return to its original shape after having experienced stresses that deformed it.

Abrasion resistance index. It is a measure that evaluates the resistance capacity of a vulcanized rubber in comparison with a specific standard under particular conditions.

Tear strength. It is defined as the minimum force required to break an inch thick sample under specified conditions.

Resistance to aging. Refers to the material's ability to resist deterioration caused by factors such as heat, light and exposure to oxygen during use or storage.

2.3. Conceptual Framework

CALICATA. The calicata is an excavation made in the ground that gives us the opportunity to investigate the structure of the soil at various depths.

GRANULAR RUBBER. This powder or fine particulate is used in a variety of applications, including the construction of sports surfaces, athletic and tennis courts, safety surfaces, acoustic insulation and is mixed with certain bitumen-derived products to improve surface characteristics, such as increasing durability and reducing noise, as reported by Donaire in 2008.

CBR. The CBR test evaluates the ability of a soil to resist shear forces under specific moisture and density conditions. ASTM refers to this test as "bearing ratio" and its standard is defined in ASTM D 1883-73.

SUB-BASE LAYER. A layer that works in conjunction with the base and serves similar purposes, it is incorporated when decreasing the thickness of the base is considered beneficial from an economic perspective, as stated by Burga in 2020.

COMPACTNESS. A typical concept for granular soils, compactness refers to the level of compaction of non-cohesive soils. Compactness is a crucial characteristic in roads, embankments and any type of fill in general, as it is directly related to the strength, deformability and stability of a fill. It is essential that the fill be adequately consolidated to prevent settlement.

MOISTURE CONTENT. This refers to the ratio between the weight of water present in a sample in its natural state and the weight of the same sample after it has been subjected to a drying process in an oven at temperatures ranging from 105ºC to 110ºC. This ratio is expressed as a percentage and can vary from a minimum value when the sample is completely dry to a maximum, which does not necessarily reach 100%.

According to Raffino (2021), this author points out the connection that exists between the quantity of matter (or of an object) and its occupied space. Basically, it is a fundamental characteristic of any substance. The maximum dry density refers to the highest value of density that a soil can achieve when compacted at ideal moisture.

GRANULOMETRY. Its purpose is to determine the gradation of particles in a soil sample, thus allowing its classification by systems such as AASHTO or USCS. This test is of great importance, since it contributes to meet the necessary requirements for the use of soils in applications such as road bases or subbases, earth dams, dikes, drainage systems and other areas where this procedure is applied.

PLASTICITY INDEX. The plasticity property is typical of fine soils, those in which the moisture content is within the range between the liquid limit and the plastic limit. In this condition, the soil can be shaped in a manner similar to plasticine or a dough, since the optimum amount of water molecules in the soil allows the forces of attraction between the clay mineral particles to be maximized.

IMPROVEMENT. Refers to construction actions carried out to improve the physical and operational condition of a pre-existing road, either to expand its capacity or to offer a higher quality service compared to its previous condition.

PLASTICITY. Moisture, a physical characteristic of a soil, is an indicator of the behavior of the soil in relation to its water content.

MODIFIED PROCTOR: A modification was made to the conventional Proctor test by increasing the compaction energy to 2,700 kN-m/m³, increasing the number of blows per layer to 56 and the number of layers to 5. In addition, the weight of the hammer (metal piston) was increased to 4.54 kg and the height from which it fell to 18 inches (45.57 cm).

RECYCLING. The process of transforming previously used materials into raw materials for the creation of new products.

STRENGTH. Resistance, a mechanical property of soil, is essential to improve the soil's ability to resist external forces.

SUBGRADE. The subgrade plays a fundamental role in supporting the loads transmitted by the pavement, giving it stability and serving as its support base. The quality of this layer directly influences the reduction of the necessary thickness of the pavement, which can lead to economic savings without compromising quality. The characteristics that must be met include a maximum strength of 3 inches, a maximum expansion of 5%, a minimum compaction degree of 95%, and a minimum thickness of 30 cm on low-traffic roads, or 50 cm on roads with an average annual daily traffic (ADAT) of more than 2000 vehicles. In addition, the subgrade plays an important role in preventing contamination of the pavement by the embankment and in preventing the pavement from being infiltrated by the earthwork materials.

SOIL. The material present in the Earth's surface layer is formed from the decomposition and fragmentation of rocks due to the weathering process, without moving away from its original location. This material is composed of three different phases: solid, liquid and gaseous, and its origin can be from both organic and inorganic sources, as pointed out in Burga's reference in 2020.

COHESIVE SOIL. A soil is considered cohesive when it is composed mainly of silt and clay, categorized as fine soils, which are characterized by strong cohesion between its particles.

CHAPTER III : HYPOTHESES AND VARIABLES

3.1 Hypotheses

3.1.1. General Hypothesis

H.G. The use of tire granular rubber influences the stabilization of the natural soil subgrade in Jr. Tupac Amaru, Tambopata District, Madre de Dios, 2022.

3.1.2. Hypothesis Specific Hypotheses

HE1: The addition of tire granular rubber will improve the Consistency Limits for subgrade stabilization on Jr. Tupac Amaru, Tambopata District, Madre de Dios, 2022.

HE.2. The addition of tire granular rubber will improve the modified Proctor for subgrade stabilization on Jr. Tupac Amaru, Tambopata District, Madre de Dios, 2022.

HE.3. The addition of tire granular rubber will improve the CBR in the stabilization of the subgrade in Jr. Tupac Amaru, Tambopata District, Madre de Dios, 2022.

3.2 Research variables

3.2.1. Independent Variable

X: Granular tire rubber

Indicators: Percentage of dosage

3.2.2. Dependent Variable

Y: Subgrade stabilization

Indicators:

- Granulometry
- Consistency limits
- Bearing Capacity (CBR)

3.3 Operationalization of variables

Variables	Dimensions	Indicators
Independent Variable: Granular rubber from tires	Dosage percentage	• Dosage percentage
Dependent variable: Subgrade stabilization	1. Granulometry	• Particle size.
	2. Consistency limits	• Liquid limit • Plastic limit • Plasticity index.
	3. CBR test	• 95% CBR • 100% CBR

Table 1: *Operationalization matrix of the variables*

CHAPTER IV: METHODOLOGICAL DESIGN

4.1 Type and design of research

4.1.1 Type of research

The current research project belongs to the field of applied research, since it is aimed at solving practical problems and providing concrete answers to specific questions. In this context, knowledge is used in a rigorous, organized and systematic way to understand reality. As part of this project, laboratory tests will be carried out that will generate results related to the research (Behar, 2008, p. 20).

4.1.2 Research design

The design of this research was characterized as experimental, since it was based on statistical analysis to test the hypothesis proposed. This type of experimental design is the only one that can establish a cause and effect relationship in one or several study groups. The objective is to determine how the use of granular rubber from tires influences the stabilization of the natural soil subgrade in Tupac Amaru Jr., Tambopata District, Madre de Dios, 2022.

Outline:

$$O_1 \rightarrow X \rightarrow O_2$$

Where:
O1 = Natural soil.
X = Granular tire rubber.
O2 = Treated soil.

4.1.3 Research level

The research level is of an explanatory nature, since its main purpose lies in revealing the relationship between one variable and another or others. These studies focus on unraveling causes and effects, which makes them research of a causal nature that requires careful control and meets other criteria of causality. Their fundamental objective is the examination of causal hypotheses and the search for the underlying causes of phenomena or events, both in the natural and social spheres. This type of research aims at identifying and analyzing independent variables, which act as possible causes, and how they influence observable dependent variables. In addition, intervening variables, which could have a secondary effect and play a relevant role in the study, are considered.

4.2. Research Method

A deductive research method has been adopted, which follows a logical process that proceeds from the most general to the most specific. It begins with a general premise or broad theory, and then employs the application of logical rules and deductive reasoning to arrive at specific and verifiable conclusions. These conclusions are tested through empirical evidence to verify if they agree with what is observed in reality.

A fundamental characteristic of the deductive method is that the resulting conclusions are necessarily true if the initial premises are true and the reasoning process is valid. However, it is crucial to consider that the validity of the conclusions depends on the quality of the initial premises and the precision of the applied logic.

This deductive approach is widely used in various scientific disciplines, such as physics, chemistry, mathematics and philosophy. It is also applied in social science research, where existing theories or conceptual frameworks are used to formulate specific hypotheses and design studies to test them.

The deductive research method is based on logical reasoning and starts from general premises or theories to derive specific conclusions.

Population and Sample

4.3.1 Population

The research does not determine the study population, since the focus is on analyzing the process of subgrade stabilization specifically in Jr. Tupac Amaru, Tambopata District, Madre de Dios, during the year 2022.

4.3.2 Sample

Three calicatas of the clayey soil located in Jr. Tupac Amaru, Tambopata District, Madre de Dios Region, during the year 2022.

4.4 Study Site

4.5 Data collection techniques

4.2.1 Techniques

Methods used to obtain data in this study will include direct observation, document review and laboratory testing.

4.2.2 Instruments

The data collection forms will be used as a tool to store the information collected.

CHAPTER V: PRESENTATION OF RESULTS

5.1. Analysis of tables and graphs

Photo N° 2. Location map of Jr. Tupac Amaru, Tambopata District, Madre de Dios.

The project of stabilization of the natural soil subgrade with the addition of granular rubber from tires in Jr. Tupac Amaru, Tambopata District, Madre de Dios, was executed in the indicated location.

Table 2

Moisture Content

DETERMINATION OF THE MOISTURE CONTENT OF THE EXTRACTED MATERIAL			
CALICATA	**1**	**2**	**3**
Weight Vessel + Natural Soil (g)	220.5	210.5	198.5
Weight Vessel + Dry Soil (g)	221.1	217.61	175.04
Container weight (g)	0	0	0
Water weight (g)	27.4	20.89	23.46
Natural Soil Weight (g)	230.1	208.5	198.5
Dry Soil Weight (g)	214.1	207.61	175.04
Moisture Content W (%)	15.00%	12.50%	13.40%
MOISTURE CONTENT	**13.63%**		

Source: Tupac Amaru, Tambopata District, Madre de Dios.

Graph N°1

Moisture content W (%).

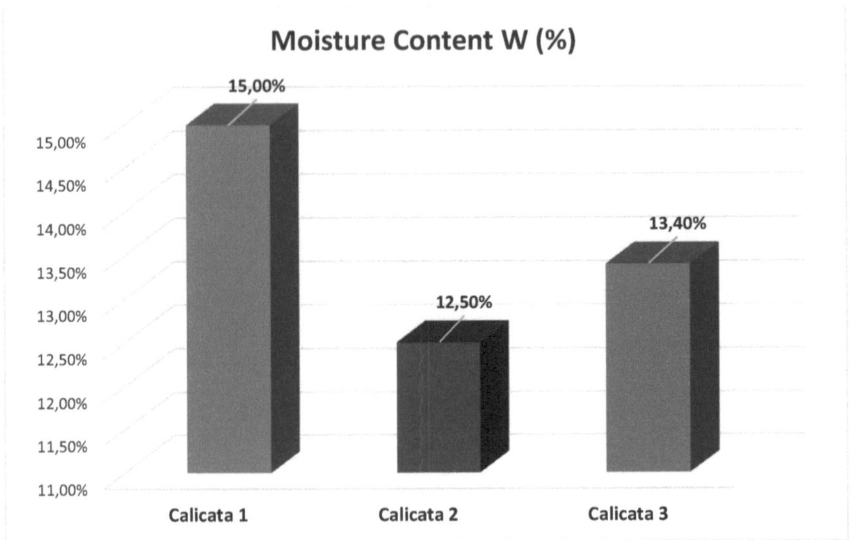

Source: Tupac Amaru, Tambopata District, Madre de Dios.

According to the data presented in Table 2 and Graph 1, it can be seen that in Jr. Tupac Amaru there is variability in moisture content among the three test pits examined. The first excavation point exhibits a moisture content of 15.00%, the second test pit shows a moisture level of 12.50%, and the third test pit presents a moisture value of 13.40%. Consequently, it can be deduced that the average moisture content in Jr. Tupac Amaru is around 13.63%.

Table 3

Liquid Limit, Plastic Limit and Index of Plasticity

PLASTIC LIMIT (LP) - ASTM 4318			
CALICATA	**1**	**2**	**3**
Capsule weight (g)	11.4	11.5	11.4
Capsule weight + Wet Soil (g)	21	20.3	21
Capsule weight + Dry Soil (g)	19.27	18.57	19.27
Dry Soil Weight (g)	7.87	7.08	8.87
Moisture Content W (%)	12.8	13	8.5
Plastic Limit (%)	**11.43%**		
LIQUID LIMIT (LL) - ASTM 4318			
CALICATA	**1**	**2**	**3**
Capsule weight (g)	37	37.2	37
Capsule weight + Wet Soil (g)	60.2	62	60.2

Capsule weight + Dry Soil (g)	54.4	59.5	54.4
Number of strokes	34	29	74
Dry soil weight (g)	16.7	19.8	16.7
Moisture Content W (%)	35.2	38.8	37
Liquid Limit (%)	**37.00%**		
PLASTICITY INDEX (IP) - ASTM 4318			
Plasticity Index (%)	**25.57%**		

Source: Calicata del Jr. Tupac Amaru, Tambopata District, Madre de Dios.

Based on the information provided in Table 3, the results of the soil consistency limits after modification in Jr. Tupac Amaru can be examined. The values obtained are as follows: the Plastic Limit (PL) is found to be 11.43%, the Liquid Limit (LL) shows a value of 37.00%, and the Plasticity Index (PI) is estimated at 25.57%.

Table 4

Modified Proctor Test

MODIFIED PROCTOR TEST						
CALICATA	**RESULTS**	**TEST 1**	**TEST 2**	**TEST 3**	**TEST 4**	**RESULT**
CALICATA 1	Compaction (Wet Density (g/cm3))	1.725	2.018	2.016	1.995	1.939
	Compaction (Dry density (g/cm3))	1.662	1.858	1.919	1.722	1.790
CALICATA 2	Compaction (Wet Density (g/cm3))	1.892	2.02	2.11	2.07	2.023
	Compaction (Dry density (g/cm3))	1.713	1.814	1.843	1.795	1.791
CALICATA 3	Compaction (Wet Density (g/cm3))	1.92	2.08	2.22	2.15	2.093
	Compaction (Dry density (g/cm3))	1.838	1.925	1.984	1.886	1.908

Source: Tupac Amaru, Tambopata District, Madre de Dios.

Analyzing the information provided in Table 4, the results of the Modified Proctor Test for the three test pits can be seen in terms of compaction. In the first test pit, a Wet

Density of 1,939 g/cm³ and a Dry Density of 1,790 g/cm³ were recorded. Regarding the second test pit, the Wet Density is 2.023 g/cm³, while the Dry Density is 1.791 g/cm³. Finally, in the third test pit, a Wet Density of 2.093 g/cm³ and a Dry Density of 1.908 g/cm³ is obtained.

HYPOTHESIS TESTING PROCESS

Significance level.

- Margin of error equivalent to 5%.

Confidence level.

- Results are 95% guaranteed

Statistician.

- ANOVA

Decision rule.

- If, Sig. < 0.05, the null hypothesis is rejected.

General Hypothesis Testing Process

H_0 : The use of granular rubber from tires does not influence the stabilization of the natural soil subgrade in Jr. Tupac Amaru, Tambopata District, Madre de Dios, 2022.

H_1 : The use of tire granular rubber influences the stabilization of the natural soil subgrade in Jr. Tupac Amaru, Tambopata District, Madre de Dios, 2022.

Table 5

ANOVA test Soil Stabilization

		Sum of squares	gl	Root mean square	F	Sig.
		ANOVA				
	Between groups	,942	4	,236	3,806	,000
Soil stabilization	Within groups	,868	14	,062		
	Total	1,810	18			

Source: Tupac Amaru, Tambopata District, Madre de Dios.

Decision:

When reviewing Table 5, it is evident that the results show a notable influence of tire granular rubber in improving the stability of the subgrade in Tupac Amaru Jr. located in the Tambopata District, Madre de Dios. This impact is confirmed with a significance value (0.000), which is less than 0.05, resulting in the rejection of the null hypothesis.

Specific Hypothesis Testing Process 1:

H_0 : The addition of tire granular rubber will not improve the Consistency Limits for subgrade stabilization on Jr. Tupac Amaru, Tambopata District, Madre de Dios, 2022.

H_1 : The addition of tire granular rubber will improve the Consistency Limits for subgrade stabilization on Jr. Tupac Amaru, Tambopata District, Madre de Dios, 2022.

Table 6

ANOVA test Consistency Boundary Improvement

		Sum of squares	gl	Root mean square	F	Sig.
Improvement of	Between groups	,876	4	,219	3,842	,000
consistency	Within groups	,792	14	,057		
limits	Total	1,668	18			

Source: Tupac Amaru, Tambopata District, Madre de Dios.

Decision:

A review of Table 6 shows that the results indicate that granular tire rubber has a significant impact on the consistency limits of the subgrade in Tupac Amaru Jr. located in the Tambopata District of Madre de Dios. This impact is supported by a significance value (0.000) that is less than 0.05, which leads to the rejection of the first specific null hypothesis.

Specific Hypothesis Testing Process 2:

H_0 : The addition of tire granular rubber will not improve the modified Proctor for subgrade stabilization on Jr. Tupac Amaru, Tambopata District, Madre de Dios, 2022.

H_1 : The addition of tire granular rubber will improve the modified Proctor for subgrade stabilization on Jr. Tupac Amaru, Tambopata District, Madre de Dios, 2022.

Table 7

ANOVA test Improvement of the Modified Proctor

ANOVA					
	Sum of squares	gl	Root mean square	F	Sig.
Modified Proctor Improvement					
Between groups	,857	4	,214	4,116	,000
Within groups	,724	14	,052		
Total	1,581	18			

Source: Tupac Amaru, Tambopata District, Madre de Dios.

Decision:

Examining Table 7, the results indicate that tire granular rubber has a substantial impact on the modified Proctor test of the subgrade in Jr. Tupac Amaru, located in the Tambopata District, Madre de Dios. This is because the significance value (0.000) is less than 0.05, which leads to the rejection of the second specific null hypothesis.

Specific Hypothesis Testing Process 3:

H_0 : The addition of tire granular rubber will not improve the CBR in subgrade stabilization on Jr. Tupac Amaru, Tambopata District, Madre de Dios, 2022.

H_1 : Addition of tire granular rubber will improve CBR in subgrade stabilization on Jr. Tupac Amaru, Tambopata District, Madre de Dios, 2022.

Table 8

ANOVA test Improvement of the CBR

		Sum of squares	gl	Root mean square	F	Sig.
		ANOVA				
CBR improvement	Between groups	,834	4	,209	4,098	,000
	Within groups	,712	14	,051		
	Total	1,552	18			

Source: Tupac Amaru, Tambopata District, Madre de Dios.

Decision:

When analyzing Table 8, the results indicate that granular tire rubber has a significant impact on the CBR (California Bearing Ratio) resistance index of the subgrade in Tupac Amaru Jr. located in the Tambopata District of Madre de Dios. This is because the significance value (0.000) is less than 0.05, which leads to the rejection of the third specific null hypothesis.

Discussion of Results

The fundamental purpose of this study was to examine how tire granular rubber affects the improvement of subgrade stability in Jr. Tupac Amaru, located in the Tambopata District of Madre de Dios, during the year 2022. In order to achieve this objective, evaluations were carried out on three essential aspects: consistency limits, modified Proctor test and bearing capacity (CBR).

Regarding the central objective of this study, the results show that the use of granular rubber from tires has a significant effect in improving the stability of the subgrade in Tupac Amaru Jr. located in the Tambopata District of Madre de Dios. This is confirmed by the significance value (0.000), which is less than 0.05, resulting in the rejection of the null hypothesis.

Regarding the first specific objective of this research, the results indicate that granular tire rubber has a significant effect on the consistency limits of the subgrade in Tupac Amaru Jr. located in the Tambopata District of Madre de Dios. This conclusion is supported by the significance value (0.000), which is less than 0.05, resulting in the rejection of the first specific null hypothesis.

Regarding the second specific objective of this study, the results indicate that granular tire rubber has a significant effect on the modified Proctor test of the subgrade in Tupac Amaru Jr. located in the Tambopata District of Madre de Dios. This impact is supported by a significance value (0.000), which is less than 0.05, leading to the rejection of the third specific null hypothesis.

In relation to the third specific objective of this study, the results show that granular tire rubber significantly influences the CBR (California Bearing Ratio) index of the subgrade in Tupac Amaru Jr. located in the Tambopata District of Madre de Dios. This effect is supported by the significance value (0.000), which is less than 0.05, leading to the rejection of the second specific null hypothesis.

The results of this study agree with the findings of Patiño, J. (2017), as they point out that classification processes of two types of soil intended for probe tests will be carried out. Experiments will be carried out using two types of probes: one of pure soil and the other of soil mixed with rubber. These tests will be executed with the purpose of evaluating the resistance through the CBR (California Bearing Ratio) test and the density through the modified Proctor test, following the standards established by ASTM, which will be the essential protocols in this study. The results obtained in the laboratory will provide the necessary information to evaluate the effectiveness of soil stabilization by means of the soil-rubber mixture, taking into account the various combinations to be carried out.

Similarly, these results are in agreement with the findings of Laica, J. (2016), who indicate that the investigation was initiated by collecting the essential materials, including the subbase and the recycled rubber polymer. In an initial phase, the physico-mechanical properties of the subbase were examined to determine its compliance with the requirements set by AASHTO and ASTM standards. The subbase was acquired from Constructora Alvarado Ortiz, located at the intersection of Arq. Lecorbusier and Socrates Streets. In addition, recycled rubber polymer was obtained from Proneumacosa, located at Km 12 of the Panamericana Norte in the "La Avelina" sector. Compaction, modified Proctor and California Bearing Ratio (CBR) tests were carried out with the incorporation of rubber in various proportions. Ultimately, the results obtained from both the sample in its natural state and the sample with the addition of rubber in different percentages were contrasted. The results showed that as the amount of rubber in our subbase increases, the resistance experiences a significant decrease.

The results are similar to Alvarez, S. (2020), where he reaches the conclusions that through various investigations carried out in different parts of the world on the use of recycled tires in soils that lack the capacity to support road structures, it has been shown that a recycled tire contains several materials that can be used to strengthen a subgrade soil with stability problems. Rubber powder, in particular, is presented as an option to reinforce unstable soils, since it can improve some of their mechanical properties, such as shear strength, cohesion and friction angle, especially in clayey soils. In addition, it has been observed that mixing 4% rubber powder with the unsuitable soil results in a significant increase in the California Bearing Ratio (CBR). In order to apply this methodology to the soils of the city of Bogotá, the properties and behavior of the materials present in the deposits of this area have been examined and characterized in detail. This has made it possible to identify the areas of the city where it would be feasible to use rubber powder to strengthen the soil. Comparison between traditional methods of soft soil improvement and the approach proposed in this report has revealed economic and environmental advantages for the city. However, due to the scarcity of research in Colombia on this particular topic, it is not possible to state with complete certainty whether rubber powder is suitable for use in the capital's soils. Nevertheless, this raises a new line of research in the country that could boost future studies related to this topic.

According to Rojas, R. (2019), the purpose of this study is to determine the impact of the introduction of recycled granular rubber from tires on soil properties. The research was conducted in a segment of Bonavista Avenue, located in the District of Carabayllo, characterized by a soil in unfavorable conditions. The methodological approach was based on the scientific method, using a quantitative approach involving data collection to address the research questions. This study is classified as applied, since theories are applied for the benefit of the study area, with an explanatory level and an experimental design that deliberately manipulated one of the variables. The population of interest was the existing soil on Bonavista Avenue, and samples were extracted from two 1.5-meter deep soil pits. However, the results of the improvement of the subgrade properties were not positive, since the incorporation of recycled granular rubber failed to improve the compaction, strength and expansion of the tested soil. Based on these findings, it is suggested to consider the use of recycled granular rubber in applications other than soil mechanics to reduce environmental contamination in more appropriate environments.

According to Cueto (2018), the results obtained in this study are consistent with his findings. The purpose of the research was to establish a technical proposal to stabilize a slope using recycled tires on the Hualituna - Curva Gervasio carriageway in the Junín Region. This work is classified as applied, since it applied theoretical knowledge to address existing problems, although an experimental design was not carried out. The population under study comprised the slopes along the carriageway, and the sample was selected from kilometers 01+570 to 01+575. The results of the investigation concluded that the technical proposal incorporating recycled tires would make possible the stabilization of the slope on the Hualituna - Curva Gervasio dirt road. In addition, the use of recycled tires with a diameter of 072 m and an inner diameter of 0.45 m, filled with sterile material composed of clayey gravel type soil, is recommended.

The results are in agreement with national research, such as that of Yucra and Huamán, R. & Muguerza, K. (2019), who took the samples to the laboratory to examine the impact of different proportions of granulated rubber (5%, 10% and 15%) on the penetration resistance when incorporated into the dry weight of the soil. In total, 12 CBR tests were conducted to evaluate the penetration resistance and determine the optimum proportion of granulated rubber. The results from test pit 3 indicated a SUCS classification of low plasticity clay (LC). An increase in CBR will be monitored at the 5% and 10% rubber percentages, but not at 15%. The optimum ratio identified was 10% granular rubber, as it raised the soil strength from 5.2% to 12.2%. On the other hand, pits 1 and 2 were classified as gravel with clay (GC) according to SUCS, and a steady decrease in CBR is evident with each increase in rubber percentage, being from 26.27% to 20.16% and from 34.06% to 28.5. %, respectively. In summary, it can be stated that granulated rubber improves cohesive soils, as long as the parameters established by the Peruvian standard MTC Suelos y Geotecnia - 2013 are met.

Similarly, the results obtained by Casimiro, V. & Melgarejo, K. (2022) present similarities with the present investigation. In this study, the use of granulated rubber as a stabilizing agent in cohesive soils was explored, evaluating its physical-mechanical properties, such as plasticity, density, moisture and bearing capacity. The methodology employed was deductive and was based on the review of more than 30 research papers related to the use of this waste tire by-product as a stabilizing agent in cohesive soils. This review provided a significant amount of information on the use of granulated rubber for the purpose of preventing future problems in road infrastructure projects. In addition, it examines how granulated rubber affects each of the properties of the cohesive soil and how these properties relate to each other. To carry out this research, soil samples were collected in the district of San Luis de Shuaro, province of Chanchamayo, Junín, which presented similar characteristics to those obtained in the review of data from various authors. The objective of this study was to analyze the behavior of cohesive soil when granulated rubber is added as a stabilizing agent. The experimental results indicated that the Shrinkage Limit increases with the incorporation of granulated rubber. The Maximum Dry Density reaches its maximum value when the soil is in its natural state, decreasing as granulated rubber is added. In relation to the CBR in critical conditions (soaked state), a decrease in resistance will be observed after the incorporation of granulated rubber. This does not optimize the physical-mechanical properties of the soil, since it does not meet the minimum criteria required to be considered a good subgrade according to the Highway Manual (MC-05-14).

CONCLUSIONS

1. It was concluded that the introduction of granular tire rubber leads to a substantial improvement in soil stabilization, since the significance value (0.000) is less than 0.05; therefore, the null hypothesis is discarded. That is, the use of tire granular rubber significantly stabilizes the soil in Jr. Tupac Amaru. In relation to the compacted samples at optimum moisture, it was observed that the inclusion of tire granular rubber resulted in a decrease in wet density and contributed to improve the CBR (California Bearing Ratio).

2. It is corroborated that consistency limits play a significant role in soil stabilization in Jr. Tupac Amaru, supported by a significance value (0.000) lower than 0.05, which results in the rejection of the first specific exception. This finding underlines the importance of consistency limits in the soil stabilization process in the mentioned locality, especially when considering compaction operations, which acquire a crucial relevance due to the specific climatic conditions of the region.

3. It is evident that the modified Proctor exerts a significant influence on soil stabilization, indicated by the significance value (0.000) which is less than 0.05. This finding implies the rejection of the second specific null hypothesis. In addition, it is important to note that the compaction operations under study become relevant in view of the climatic conditions of the region. This aspect further highlights the importance of understanding and considering soil consistency limits for effective stabilization.

4. It is concluded that the influence of CBR on soil stabilization in Jr. Tupac Amaru is significant, supported by a significance value (0.000) lower than 0.05, resulting in the rejection of the third specific null hypothesis. Furthermore, it is relevant to highlight that the soil mixtures stabilized with tire granular rubber exhibited remarkable improvements, such as increased cohesion, reduced expansion and contraction, as well as increased bearing capacity. These results underscore the effectiveness and usefulness of tire granular rubber in improving the geotechnical properties of soil.

RECOMMENDATIONS

1. It is recommended to use 15% CC for an optimum mix because it achieves a considerable increase in CBR in relation to natural soil of 10.30%.

2. It is recommended to analyze coal ash to improve mechanical properties in other types of soils.

3. It is recommended to carry out studies to improve mechanical properties in other pavement layers such as base or sub-base, since coal ash has appropriate components for stabilization.

4. It is always advisable to have the guidance and supervision of a geotechnical engineer or a professional with experience in soil stabilization projects with enzymatic products. Their knowledge and experience are critical to the success of the project.

Álvarez, N. & Gutiérrez, J. (2019). Experimental study of the mechanical effect of a clayey soil by adding rubber powder for geotechnical applications [Universidad Peruana de Ciencias Aplicadas].
https://repositorioacademico.upc.edu.pe/handle/10757/648723

Arias Gonzáles, J. L. (2020). Techniques and instruments of scientific research. http://repositorio.concytec.gob.pe/handle/20.500.12390/2238

Casimiro, V. & Melgarejo, K. (2022). Cohesive soil stabilized with Granulated Rubber to improve the physical-mechanical properties of a subgrade in rural areas.

Castro, A. (2017). Stabilization of clay soils with rice husk ash for subgrade improvement.

Cuipal, B. (2018). Stabilization of clayey soil subgrade using synthetic polymer on the Chachapoyas-Huancas road, Amazonas.

Delgado, C. & Mormontoy, V. (2021). Stabilization of Clay Soils with the addition of Corn Cob Ash and Lime (Undergraduate Thesis). Universidad Andina del Cusco. https://repositorio.uandina.edu.pe/handle/20.500.12557/4587

Díaz Vásquez, F. (2018). Subgrade improvement using rice husk ash on the road Dv San Martin - Lonya Grande, Amazonas 2018. http://repositorio.ucv.edu.pe/bitstream/handle/UCV/25951/Díaz_VF.pdf?sequence=1 &isAllowed=y

Diaz, K. & Torres, R. (2019). Incorporation of Tire rubber particles to improve mechanical properties in clayey soils. http://repositorio.unj.edu.pe/handle/UNJ/236

Laica, J. (2016). Influence of the inclusion of recycled polymer (rubber) on the mechanical properties of a subbase [Universidad Técnica de Ambato]. In American Journal of Orthodontics and Dentofacial Orthopedics (Vol. 20, Issue 1). https://repositorio.uta.edu.ec/handle/123456789/24440

Garibay, S. A. T. (2018). Geology and Geotechnics. In Geología y Geotecnia (p. 28). https://www.fceia.unr.edu.ar/geologiaygeotecnia/TIPOS DE SUELO.pdf.

Gutiérrez, F. & Rojas, H. (2020). Influence of the addition of granular rubber on the mechanical characteristics of the subgrade in cohesive soils, Lima-2020. https://hdl.handle.net/20.500.12692/76304

Huamán, R. & Muguerza, K. (2019). Influence of granulated rubber in cohesive soils related to the penetration resistance property (CBR), 2019. https://hdl.handle.net/20.500.12692/44767

Martínez, R. (2021). Subgrade stabilization incorporating rubber and lime, in Chimpu Ocllo Ave., Carabayllo, 2020. https://hdl.handle.net/20.500.12692/69294

MTC (2018). Glossary of frequently used terms in road infrastructure projects. Portal Del MTC, 27. http://transparencia.mtc.gob.pe/idm_docs/normas_legales/1_0_4032.pdf

Patiño, J. (2017). Soil stabilization using recycled rubber additions. http://repositorio.ucsg.edu.ec/handle/3317/9159

Rivera, J., Aguirre-Guerrero, A., Mejía de Gutiérrez, R., & Orobio, A. (2020). Chemical stabilization of soils - Conventional and alkaline activated materials.

Ravichandran, PT, Prasad, AS, Krishnan, KD & Rajkumar, PRK (2016). Effect of addition of shredded rubber from scrap tires on weak soil stabilization. Indian Journal of Science and Technology. https://doi.org/10.17485/ijst/2016/v9i5/87259

Rodríguez, D. (2021). Incorporation of granulated rubber to improve the physical and mechanical behavior in the subgrade of clayey soils, Puno 2021. https://repositorio.ucv.edu.pe/handle/20.500.12692/73080

Rojas, R. (2019). Subgrade improvement incorporating recycled granular rubber in Bonavista Avenue, Carabayllo, Lima - 2019.

ANNEXES

Annex 1: Consistency matrix

Title: STABILIZATION OF THE NATURAL SOIL SUBGRADE WITH ADDITION OF GRANULAR TIRES RUBBER IN JR. TUPAC AMARU, TAMBOPATA DISTRICT, MADRE DE DIOS, 2022.				
PROBLEM	**OBJECTIVE**	**HYPOTHESIS**	**VARIABLES**	**METHODOLOGY**
GENERAL PROBLEM To what extent does the use of tire granular rubber influence the stabilization of the natural soil subgrade in Jr. Tupac Amaru, Tambopata District, Madre de Dios, 2022? **SPECIFIC PROBLEMS** **PE1:** How will the addition of granular rubber improve the Consistency Limits for subgrade stabilization on Jr. Tupac Amaru, Tambopata District, Madre de Dios, 2022? **PE2:** How will the addition of granular rubber improve the modified Proctor for subgrade stabilization on Jr. Tupac Amaru, Tambopata District, Madre de Dios, 2022? To what extent will the addition of granular rubber improve the CBR in the stabilization of the subgrade in Jr. Tupac Amaru, Tambopata District, Madre de Dios, 2022?	**GENERAL OBJECTIVE** To determine how the use of granular rubber from tires influences the stabilization of the natural soil subgrade in Jr. Tupac Amaru, Tambopata District, Madre de Dios, 2022. **SPECIFIC OBJECTIVES** **SO1:** Determine how the addition of granular rubber will improve the Consistency Limits for subgrade stabilization in Jr. Tupac Amaru, Tambopata District, Madre de Dios, 2022. **SO2:** Determine how the addition of granular rubber will improve the modified Proctor for subgrade stabilization in Jr. Tupac Amaru, Tambopata District, Madre de Dios, 2022. **SO3:** Determine how the addition of granular rubber will improve the CBR in the stabilization of the subgrade in Jr. Tupac Amaru, Tambopata District, Madre de Dios, 2022.	**GENERAL HYPOTHESIS** The use of tire granular rubber influences the stabilization of the natural soil subgrade in Jr. Tupac Amaru, Tambopata District, Madre de Dios, 2022. **SPECIFIC ASSUMPTIONS** **HE1:** The addition of granular rubber will improve the Consistency Limits for subgrade stabilization on Jr. Tupac Amaru, Tambopata District, Madre de Dios, 2022. **HE.2.** The addition of granular rubber will improve the modified Proctor for subgrade stabilization on Jr. Tupac Amaru, Tambopata District, Madre de Dios, 2022. **HE.3.** The addition of granular rubber will improve the CBR in the stabilization of the subgrade in Jr. Tupac Amaru, Tambopata District, Madre de Dios, 2022.	**INDEPENDENT VARIABLE:** Granular rubber from tires **DEPENDENT VARIABLE:** Subgrade stabilization **DIMENSIONS:** • Granulometry • Consistency limits • Bearing Capacity (CBR)	DESIGN: - Experimental RESEARCH TYPE: - Applied RESEARCH LEVEL: - Explanatory POPULATION: Not applicable. SAMPLE: 3 calicatas. TECHNICAL: Observation. INSTRUMENT: Observation guide. INTERPRETATION OF RESULTS Descriptive and inferential statistics

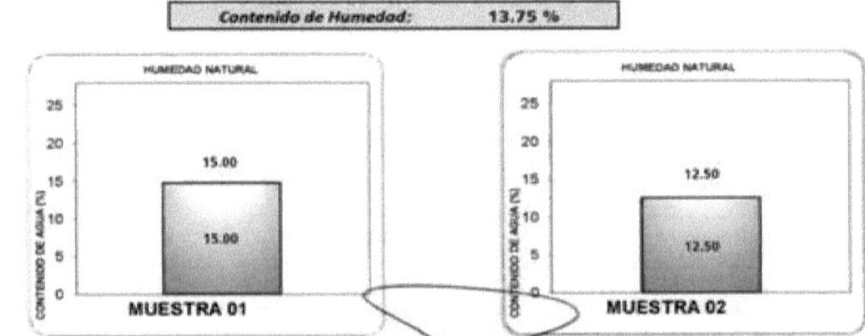

GEOIN GEOTECNIA E INGENIEROS EIRL.

CONTENIDO DE HUMEDAD (ASTM D2216-19, NTP 339.127)

Datos del poyecto

Proyecto	:	"ESTABILIZACIÓN DE LA SUBRASANTE DEL SUELO NATURAL CON ADICIÓN DE CAUCHO GRANULAR DE NEUMÁTICOS EN EL JR. TUPAC AMARU, DISTRITO TAMBOPATA, MADRE DE DIOS, 2022"
Lugar	:	Jr. TUPAC AMARU, DISTRITO DE TAMBOPATA, MADRE DE DIOS
Dist/Prov.	:	TAMBOPATA – TAMBOPATA
Solicitante	:	EDWARD JIMMY PANDIA YAÑEZ
Hecho por	:	ING. VICTOR HUGO CARAZAS MAYANGA
Fecha	:	20/06/2022

Datos de la Muestra | **Datos del Equipo Calibrado**

Calicata	:	P-1	Equipo :
Profundidad	:	1.50 m.	HORNO DIGITAL de 0°C a 300°C
condicion	:	Alterada	Certificado de Calibración N° : MT-LT-050-2020 del 02/12/2020

Datos y resultados de ensayo

CONTENIDO DE HUMEDAD

N° de Capsula		M - 01	M-02
Peso Recipiente + Suelo Natural	g	220.50	210.50
Peso Recipiente + Suelo Seco	g	221.10	217.61
Peso Recipiente	g	0.00	0.00
Peso del agua	g	27.80	20.89
Peso del Suelo Natural	g	230.10	208.50
Peso del Suelo Seco	g	214.10	207.61
Contenido de Humedad (w)	%	15.00	12.50

Contenido de Humedad:	13.75 %

GEOTECNIA E INGENIEROS E.I.R.L.

VICTOR HUGO CARAZAS MAYANGA
INGENIERO CIVIL
CIP: 108192
AREA DE GEOTECNIA

GEOIN GEOTECNIA E INGENIEROS EIRL.

LABORATORIO DE MECANICA DE SUELOS · CONCRETO Y MATERIALES · ESTUDIOS GEOTECNICOS JANPLOS Y ROCAS) · CONTROL DE CALIDAD DE OBRAS CIVILES
CONSULTORIA ESPECIALIZADA · PERFORACION Y SONDAJE PARA ACUIFEROS Y CIMENTACIONES PROFUNDAS · HINCADO DE PILOTES · PROSPECCION GEOFISICA
PUERTO MALDONADO JR. CUSCO 128 - TAMBOPATA · CUSCO URB. MEZA REDONDA A-9 - CUSCO · 982757087 · 082-574754 · RUC : 20480001981

CONTENIDO DE HUMEDAD (ASTM D2216-19, NTP 339.127)

Datos del poyecto

Proyecto	:	"ESTABILIZACIÓN DE LA SUBRASANTE DEL SUELO NATURAL CON ADICIÓN DE CAUCHO GRANULAR DE NEUMÁTICOS EN EL JR. TUPAC AMARU, DISTRITO TAMBOPATA, MADRE DE DIOS, 2022"
Lugar	:	Jr. TUPAC AMARU, DISTRITO DE TAMBOPATA, MADRE DE DIOS
Dist/Prov.	:	TAMBOPATA – TAMBOPATA
Solicitante	:	EDWARD JIMMY PANDIA YAÑEZ
Hecho por	:	ING. VICTOR HUGO CARAZAS MAYANGA
Fecha	:	20/06/2022

Datos de la Muestra

			Datos del Equipo Calibrado
			Equipo :
Calicata	:	P-3	HORNO DIGITAL de 0°C a 300°C
Profundidad	:	1.50 m.	Certificado de Calibración N° :
condicion	:	Alterada	MT-LT-050-2020 del 02/12/2020

Datos y resultados de ensayo

CONTENIDO DE HUMEDAD

N° de Capsula		M - 01	M-02
Peso Recipiente + Suelo Natural	g	198.50	201.20
Peso Recipiente + Suelo Seco	g	175.04	177.30
Peso Recipiente	g	0.00	0.00
Peso del agua	g	23.46	23.90
Peso del Suelo Natural	g	198.50	201.20
Peso del Suelo Seco	g	175.04	177.30
Contenido de Humedad (w)	%	13.40	13.48

Contenido de Humedad:	13.44 %

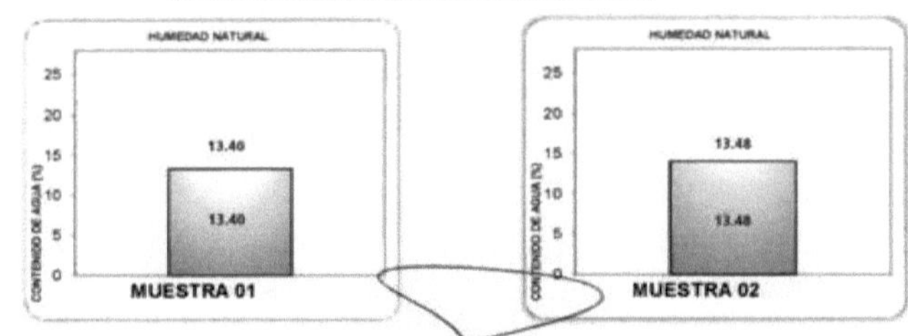

GEOTECNIA E INGENIEROS E.I.R.L.

VICTOR HUGO CARAZAS MAYANGA
INGENIERO CIVIL
CIP : 108352
AREA DE GEOTECNIA

CONTENIDO DE HUMEDAD (ASTM D2216-19, NTP 339.127)

Datos del poyecto

Proyecto	:	"ESTABILIZACIÓN DE LA SUBRASANTE DEL SUELO NATURAL CON ADICIÓN DE CAUCHO GRANULAR DE NEUMÁTICOS EN EL JR. TUPAC AMARU, DISTRITO TAMBOPATA, MADRE DE DIOS, 2022"
Lugar	:	Jr. TUPAC AMARU, DISTRITO DE TAMBOPATA, MADRE DE DIOS
Dist/Prov.	:	TAMBOPATA – TAMBOPATA
Solicitante	:	EDWARD JIMMY PANDIA YAÑEZ
Hecho por	:	ING. VICTOR HUGO CARAZAS MAYANGA
Fecha	:	20/06/2022

Datos de la Muestra

Datos del Equipo Calibrado

Calicata	:	P-1	Equipo :
Profundidad	:	1.50 m.	HORNO DIGITAL de 0°C a 300°C
condicion	:	Alterada	Certificado de Calibración N° :
			MT-LT-050-2020 del 02/12/2020

Datos y resultados de ensayo

CONTENIDO DE HUMEDAD

N° de Capsula		M - 01	M-02
Peso Recipiente + Suelo Natural	g	240.50	268.50
Peso Recipiente + Suelo Seco	g	241.10	237.61
Peso Recipiente	g	0.00	0.00
Peso del agua	g	27.40	30.89
Peso del Suelo Natural	g	240.10	268.50
Peso del Suelo Seco	g	214.10	237.61
Contenido de Humedad (w)	%	12.80	13.00

Contenido de Humedad:	12.90 %

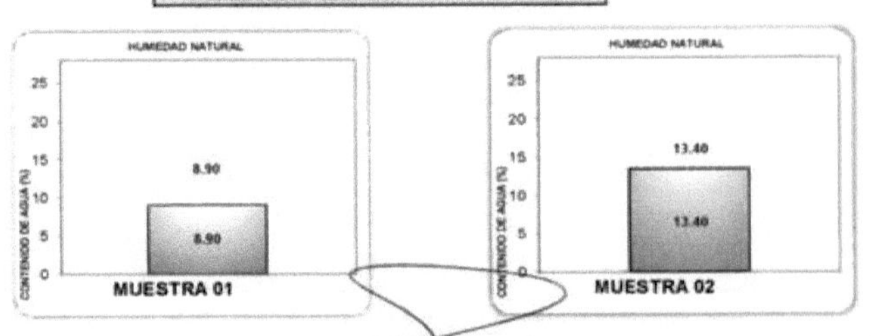

GEOTECNIA E INGENIEROS E.I.R.L.

VICTOR HUGO CARAZAS MAYANGA
INGENIERO CIVIL
CIP : 108 352
AREA DE GEOTECNIA

LABORATORIO DE MECANICA DE SUELOS · CONCRETO Y MATERIALES · ESTUDIOS GEOTECNICOS (SUELOS Y ROCAS) · CONTROL DE CALIDAD DE OBRAS CIVILES
CONSULTORIA ESPECIALIZADA · PERFORACION Y SONDAJE PARA ACUIFEROS Y CIMENTACIONES PROFUNDAS · HINCADO DE PILOTES · PROSPECCION GEOFISICA
PUERTO MALDONADO JR. CUSCO 138 - TAMBOPATA   CUSCO URB. MEZA REDONDA A-9 - CUSCO   982737067   982-574794   RUC : 20480531861

LIMITES DE CONSISTENCIA (ASTM D4318, NTP 339.129)

Datos del poyecto

Proyecto	:	"ESTABILIZACIÓN DE LA SUBRASANTE DEL SUELO NATURAL CON ADICIÓN DE CAUCHO GRANULAR DE NEUMÁTICOS EN EL JR. TUPAC AMARU, DISTRITO TAMBOPATA, MADRE DE DIOS, 2022"
Lugar	:	Jr. TUPAC AMARU, DISTRITO DE TAMBOPATA, MADRE DE DIOS
Dist/Prov.	:	TAMBOPATA – TAMBOPATA
Solicitante	:	EDWARD JIMMY PANDIA YAÑEZ
Hecho por	:	ING. VICTOR HUGO CARAZAS MAYANGA
Fecha	:	20/05/2022

Datos de la Muestra / Datos del Equipo Calibrado

Datos de la Muestra			Datos del Equipo Calibrado
Calicata	:	P-1	Equipo : CAZUELA DE CASAGRANDE
Profundidad	:	1.50 m.	Certificado de Calibración N° :
condicion	:	Alterada	MT-LT-137-2020 del 02/12/2020

Datos y resultados de ensayo

LIMITE PLASTICO - ASTM D 4318   LP (%) = 12.90

Muestra	1	2
Numero de capsula	49.00	98.00
Peso de la Capsula (g)	11.40	11.50
Peso de la Capsula+Suelo Humedo (g)	21.00	20.30
Peso de la Capsula+ Suelo Seco (g)	19.27	18.57
Peso del Suelo Seco (g)	7.87	7.08
Contenido de Humedad (w)	12.80	13.00

LIMITE LIQUIDO - ASTM D 4318   LL (%) = 38.63   IP (%) 15.40

Muestra	A	B	C
Numero de capsula	54.00	20.00	27.00
Peso de la Capsula (g)	37.00	37.20	39.80
Peso de la Capsula+Suelo Humedo (g)	60.20	62.00	59.00
Peso de la Capsula+ Suelo Seco (g)	54.40	59.50	55.45
Numero de golpes	34.00	29.00	21.00
Peso del Suelo Seco (g)	16.70	19.80	15.80
Contenido de Humedad (w)	35.20	38.80	41.90

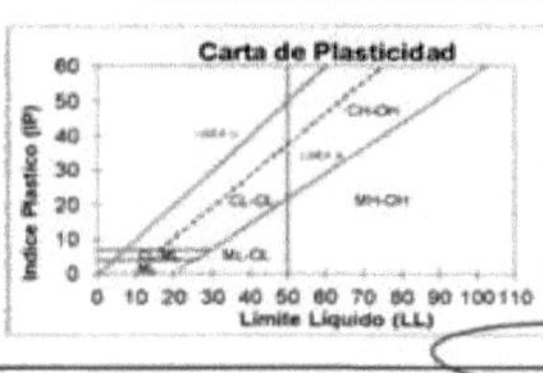

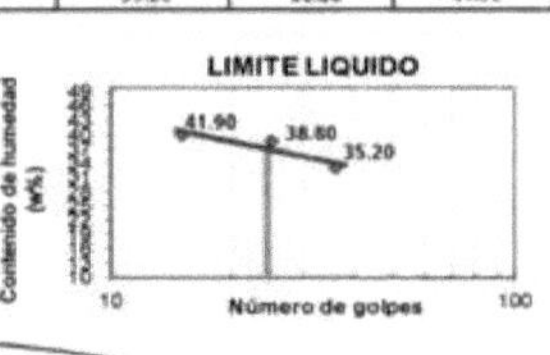

GEOTECNIA E INGENIEROS E.I.R.L.
VICTOR HUGO CARAZAS MAYANGA
INGENIERO CIVIL
CIP - 08132
AREA DE GEOTECNIA

LIMITES DE CONSISTENCIA   (ASTM D4318, NTP 339.129)

Datos del poyecto

Proyecto	:	"ESTABILIZACIÓN DE LA SUBRASANTE DEL SUELO NATURAL CON ADICIÓN DE CAUCHO GRANULAR DE NEUMÁTICOS EN EL JR. TUPAC AMARU, DISTRITO TAMBOPATA, MADRE DE DIOS, 2022"
Lugar	:	Jr. TUPAC AMARU, DISTRITO DE TAMBOPATA, MADRE DE DIOS
Dist/Prov.	:	TAMBOPATA – TAMBOPATA
Solicitante	:	EDWARD JIMMY PANDIA YAÑEZ
Hecho por	:	ING. VICTOR HUGO CARAZAS MAYANGA
Fecha	:	20/06/2022

Datos de la Muestra

			Datos del Equipo Calibrado	
Calicata	:	P-1	Equipo	: CAZUELA DE CASAGRANDE
Profundidad	:	1.50 m.	Certificado de Calibración N°	:
condicion	:	Alterada	MT-LT-137-2020 del 02/12/2020	

Datos y resultados de ensayo

LIMITE PLASTICO - ASTM D 4318	LP (%) =	9.00
Muestra	1	2
Numero de capsula	174.00	128.00
Peso de la Capsula (g)	11.40	11.50
Peso de la Capsula+Suelo Humedo (g)	21.00	20.30
Peso de la Capsula+ Suelo Seco (g)	19.27	18.57
Peso del Suelo Seco (g)	8.87	8.08
Contenido de Humedad (w)	8.50	9.50

LIMITE LIQUIDO - ASTM D 4318	LL (%) =	34.86	IP (%) 13.20
Muestra	A	B	C
Numero de capsula	91.00	66.00	41.00
Peso de la Capsula (g)	37.00	37.20	39.80
Peso de la Capsula+Suelo Humedo (g)	60.20	62.00	59.00
Peso de la Capsula+ Suelo Seco (g)	54.40	59.50	55.45
Numero de golpes	74.00	67.00	62.00
Peso del Suelo Seco (g)	16.70	19.80	15.80
Contenido de Humedad (w)	37.00	35.10	32.50

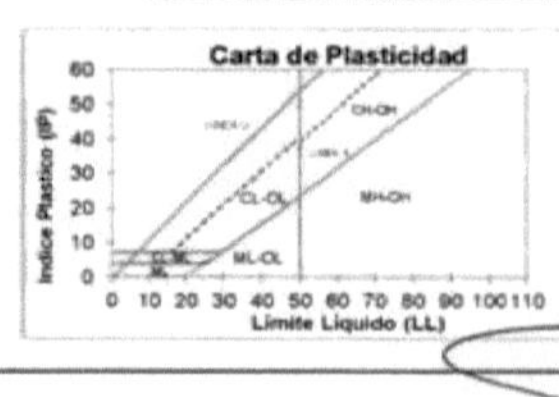

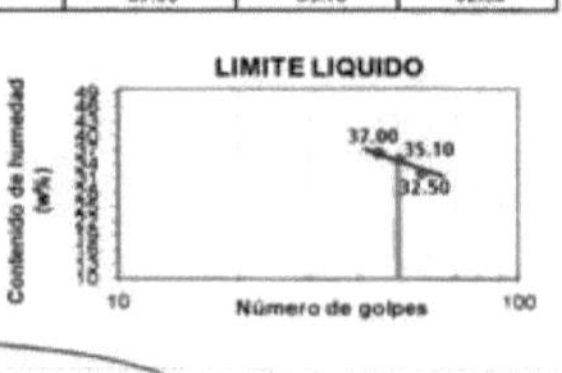

GEOTECNIA E INGENIEROS E.I.R.L.

VICTOR HUGO CARAZAS MAYANGA
INGENIERO CIVIL
CIP N° 108152
AREA DE GEOTECNIA

LIMITES DE CONSISTENCIA   (ASTM D4318, NTP 339.129)

Datos del poyecto

Proyecto	:	"ESTABILIZACIÓN DE LA SUBRASANTE DEL SUELO NATURAL CON ADICIÓN DE CAUCHO GRANULAR DE NEUMÁTICOS EN EL JR. TUPAC AMARU, DISTRITO TAMBOPATA, MADRE DE DSOS, 2022"
Lugar	:	Jr. TUPAC AMARU, DISTRITO DE TAMBOPATA, MADRE DE DIOS
Dist/Prov.	:	TAMBOPATA – TAMBOPATA
Solicitante	:	EDWARD JIMMY PANDIA YAÑEZ
Hecho por	:	ING. VICTOR HUGO CARAZAS MAYANGA
Fecha	:	20/06/2022

Datos de la Muestra / Datos del Equipo Calibrado

Calicata	:	P-1	Equipo :	CAZUELA DE CASAGRANDE
Profundidad	:	1.50 m.	Certificado de Calibración N° :	
condicion	:	Alterada	MT-LT-137-2020 del 02/12/2020	

Datos y resultados de ensayo

LIMITE PLASTICO - ASTM D 4318	LP (%) =	20.00
Muestra	**1**	**2**
Numero de capsula	81.00	188.00
Peso de la Capsula (g)	11.40	11.50
Peso de la Capsula+Suelo Humedo (g)	21.00	20.30
Peso de la Capsula+ Suelo Seco (g)	19.27	18.57
Peso del Suelo Seco (g)	19.87	18.08
Contenido de Humedad (w)	25.20	19.80

LIMITE LIQUIDO - ASTM D 4318	LL (%) =	34.93	IP (%) 14.93
Muestra	**A**	**B**	**C**
Numero de capsula	66.00	36.00	55.00
Peso de la Capsula (g)	37.00	37.20	39.80
Peso de la Capsula+Suelo Humedo (g)	60.20	62.00	59.00
Peso de la Capsula+ Suelo Seco (g)	54.40	59.50	55.45
Numero de golpes	34.00	24.00	19.00
Peso del Suelo Seco (g)	16.70	19.80	15.80
Contenido de Humedad (w)	36.80	35.40	33.20

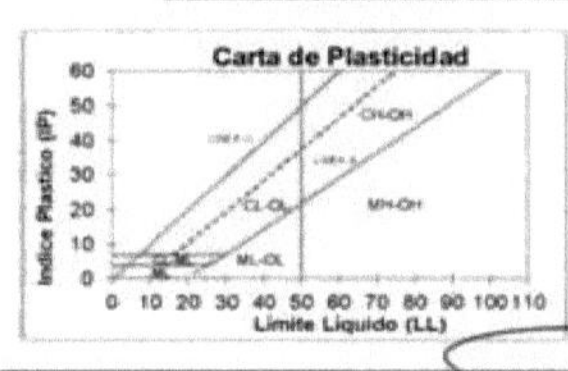

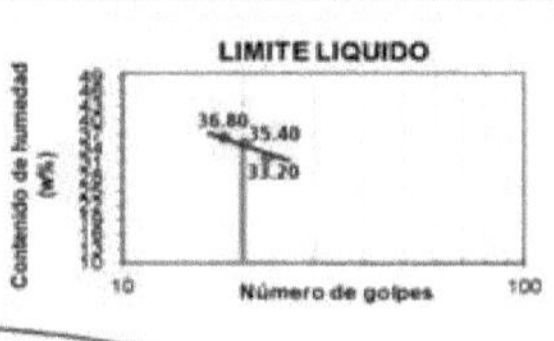

GEOTECNIA E INGENIEROS E.I.R.L.

VICTOR HUGO CARAZAS MAYANGA
INGENIERO CIVIL
CIP: 108352
AREA DE GEOTECNIA

GEOIN GEOTECNIA E INGENIEROS EIRL.

ENSAYO PROCTOR MODIFICADO (ASTM D1557-12, NTP 339.142)

Datos del poyecto

Proyecto	:	"ESTABILIZACIÓN DE LA SUBRASANTE DEL SUELO NATURAL CON ADICIÓN DE CAUCHO GRANULAR DE NEUMÁTICOS EN EL JR. TUPAC AMARU, DISTRITO TAMBOPATA, MADRE DE DIOS, 2022"
Lugar	:	Jr. TUPAC AMARU, DISTRITO TAMBOPATA, MADRE DE DIOS
Solicitante	:	EDWARD JIMMY PANDIA YAÑEZ

Dist/Prov. TAMBOPATA – TAMBOPATA
Hecho por ING. VICTOR HUGO CARAZAS MAYANGA

Datos de la Muestra

Datos del Equipo Calibrado

			Fecha	:	20/06/2022
Colicata	:	P-1			
Profundidad	:	1.50 m.			
condicion	:	Alterada			

Equipo :
PISÓN MANUAL DE PROCTOR MOD.
Certificado de Calibración N° :
MT-IV-141-2020 del 02/28/2020

Datos y resultados de ensayo

Compactacion	Codigo de molde : P1		Metodo : A molde de 4"	
Prueba N°	1	2	3	4
Numero de capas	5	5	5	5
Numero de golpes	36	36	36	36
Peso suelo + molde (g)	6120	6340	6434	6320
Peso del molde (g)	6057	6058	6057	6057
Peso del suelo humedo compactado (g)	1676	1895	1990	1876
Volumen del molde (cm³)	940.45	940.5	940.05	940.5
Densidad húmeda (g/cm³)	1.725	2.018	2.016	1.995
Humedad				
N° de tara	163	233	231	154
Tara + Suelo Humedo (g)	501.20	500.20	532.20	511.40
Tara + Suelo Seco (g)	470.06	463.57	486.26	456.38
Peso de la tara	37.60	37.60	37.36	37.60
Peso del agua	31.14	36.63	45.94	53.20
Peso de suelo seco (g)	432.46	425.97	448.63	420.87
Humedad (%)	8.80	9.86	10.24	12.60
Densidad Seca (g/cm³)	1.662	1.858	1.919	1.772

Maxima Densidad Seca (g/cm³) : 1.926 **Optimo Contenido de Humedad (%):** 10.315

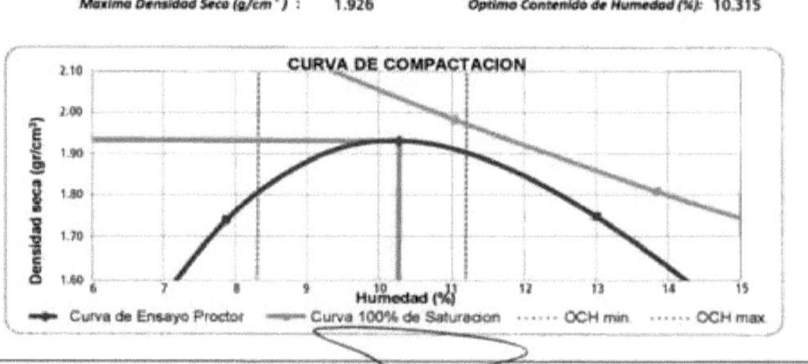

GEOTECNIA E INGENIEROS E.I.R.L.

VICTOR HUGO CARAZAS MAYANGA
INGENIERO CIVIL
CIP N° 54-152
AREA DE GEOTECNIA

ENSAYO PROCTOR MODIFICADO (ASTM D1557-12, NTP 339.142)

Datos del poyecto

Proyecto	:	"ESTABILIZACIÓN DE LA SUBRASANTE DEL SUELO NATURAL CON ADICIÓN DE CAUCHO GRANULAR DE NEUMÁTICOS EN EL JR. TUPAC AMARU, DISTRITO TAMBOPATA, MADRE DE DIOS, 2022"		
Lugar	:	Jr. TUPAC AMARU, DISTRITO DE TAMBOPATA, MADRE DE DIOS	Dist/Prov.	TAMBOPATA – TAMBOPATA
Solicitante	:	EDWARD JIMMY PANDIA YAÑEZ	Hecho por	ING. VICTOR HUGO CARAZAS MAYANGA

Datos de la Muestra

Datos del Equipo Calibrado

			Fecha :	20/06/2022	Equipo :
Calicata	:	P-1			PISÓN MANUAL DE PROCTOR MOD.
Profundidad	:	1.50 m.			Certificado de Calibración N° :
condicion	:	Alterada			MT-IV-141-2020 del 02/28/2020

Datos y resultados de ensayo

Compactacion	Codigo de molde : P1	Metodo : A molde de 4"		
Prueba N°	**1**	**2**	**3**	**4**
Numero de capas	5	5	5	5
Numero de golpes	56	56	56	56
Peso suelo + molde (g)	10000	10290	10468	10392
Peso del molde (g)	5997	5997	5997	5997
Peso del suelo humedo compactado (g)	4003	4293	4471	4395
Volumen del molde (cm³)	2122.0	2122.0	2122.0	2122.0
Densidad húmeda (g/cm³)	1.892	2.020	2.110	2.070
Humedad				
N° de tara	317	310	270	235
Tara + Suelo Humedo (g)	190.20	193.20	174.00	182.20
Tara + Suelo Seco (g)	180.21	176.11	155.10	158.58
Peso de la tara	19.60	17.60	37.59	19.65
Peso del agua	10.99	17.49	18.65	24.62
Peso de suelo seco (g)	160.52	150.39	138.39	138.21
Humedad (%)	8.86	10.41	14.03	11.70
Densidad Seca (g/cm³)	1.713	1.814	1.843	1.795

Maxima Densidad Seca (g/cm³) :   1.818   **Optimo Contenido de Humedad (%):** 12.22

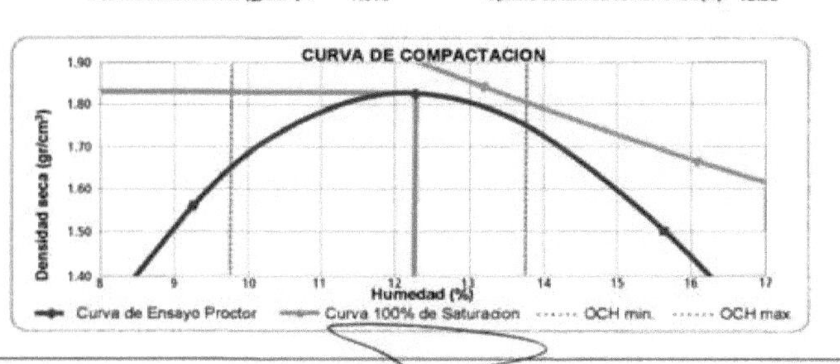

GEOTECNIA E INGENIEROS E.I.R.L.

VICTOR HUGO CARAZAS MAYANGA
INGENIERO CIVIL
CIP N° 90152
AREA DE GEOTECNIA

GEOIN GEOTECNIA E INGENIEROS EIRL.
LABORATORIO DE MECÁNICA DE SUELOS - CONCRETO Y MATERIALES - ESTUDIOS GEOTÉCNICOS (SUELOS Y ROCAS) - CONTROL DE CALIDAD DE OBRAS CIVILES
CONSULTORÍA ESPECIALIZADA - PERFORACIÓN Y SONDAJE PARA ECUIVIENDA Y CIMENTACIONES PROFUNDAS - HINCADO DE PILOTES - PROSPECCIÓN SÍSMICA
▪ PUERTO MALDONADO JR. CUSCO 132 - TAMBOPATA ▪ CUSCO URB. MEZA REDONDA A-8 - CUSCO 982737067 M 082-574754 RUC : 20490631991

ENSAYO PROCTOR MODIFICADO (ASTM D1557-12, NTP 339.142)

Datos del poyecto

Proyecto	:	"ESTABILIZACIÓN DE LA SUBRASANTE DEL SUELO NATURAL CON ADICIÓN DE CAUCHO GRANULAR DE NEUMÁTICOS EN EL JR. TUPAC AMARU, DISTRITO TAMBOPATA, MADRE DE DIOS, 2022"

Lugar	:	Jr. TUPAC AMARU, DISTRITO TAMBOPATA, MADRE DE DIOS	Dist/Prov.	TAMBOPATA — TAMBOPATA
Solicitante	:	EDWARD JIMMY PANDIA YAÑEZ	Hecho por	ING. VICTOR HUGO CARAZAS MAYANGA

Datos de la Muestra

Datos del Equipo Calibrado

			Fecha : 20/06/2022	Equipo :
				PISÓN MANUAL DE PROCTOR MOD.
Calicata	:	P-1		
Profundidad	:	1.50 m.		Certificado de Calibración N° :
condicion	:	Alterada		MT-IV-141-2020 del 02/28/2020

Datos y resultados de ensayo

Compactacion	Codigo de molde : P1		Metodo : A molde de 4"	
Prueba N°	**1**	**2**	**3**	**4**
Numero de capas	5	5	5	5
Numero de golpes	56	56	56	56
Peso suelo + molde (g)	10071	10406	10697	10566
Peso del molde (g)	5997	5997	5997	5997
Peso del suelo humedo compactado (g)	4074	4409	4700	4569
Volumen del molde (cm³)	2120.0	2120.0	2120.0	2120.0
Densidad húmeda (g/cm³)	1.920	2.080	2.220	2.150
Humedad				
N° de tara	**270**	**380**	**90**	**220**
Tara + Suelo Humedo (g)	179.40	163.60	142.40	121.20
Tara + Suelo Seco (g)	171.50	153.00	130.40	109.50
Peso de la tara	18.90	18.90	26.00	26.00
Peso del agua	7.90	10.60	12.00	11.70
Peso de suelo seco (g)	152.60	134.10	104.40	83.50
Humedad (%)	4.53	8.03	11.75	14.25
Densidad Seca (g/cm³)	1.838	1.925	1.984	1.886

Maxima Densidad Seca (g/cm³) : 1.985   **Optimo Contenido de Humedad (%):** 11.305

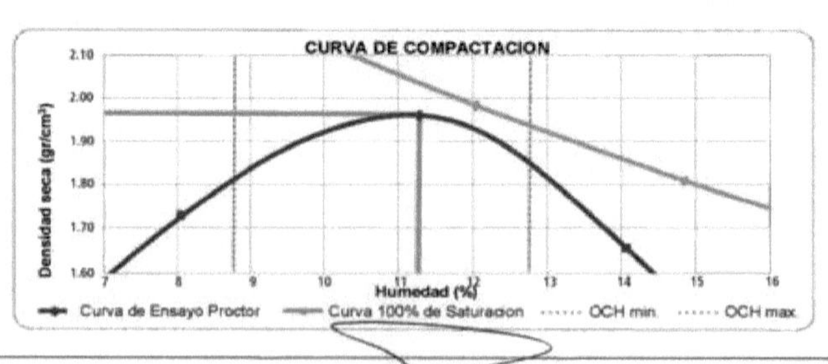

GEOTECNIA E INGENIEROS E.I.R.L.

VICTOR HUGO CARAZAS MAYANGA
INGENIERO CIVIL
CIP 108152
ÁREA DE GEOTECNIA

GEOIN GEOTECNIA E INGENIEROS EIRL.
LABORATORIO DE MECANICA DE SUELOS - CONCRETO Y MATERIALES - ESTUDIOS GEOTECNICOS (SUELOS Y ROCAS) - CONTROL DE CALIDAD DE OBRAS CIVILES
CONSULTORIA ESPECIALIZADA - PERFORACION Y BOMBAS PARA ACUIFEROS Y CIMENTACIONES PROFUNDAS - HINCADO DE PILOTES - PROSPECCION GEOFISICA
PUERTO MALDONADO JR. CUSCO 138 - TAMBOPATA   CUSCO URB. MEZA REDONDA A-9 - CUSCO   082737987   083-574754   RUC : 20490031061

VALOR RELATIVO DE SOPORTE CBR (ASTM D1883-16, NTP 339.145)

Datos del poyecto

Proyecto	"ESTABILIZACIÓN DE LA SUBRASANTE DEL SUELO NATURAL, CON ADICIÓN DE CAUCHO GRANULAR DE NEUMÁTICOS EN EL JR. TUPAC AMARU, DISTRITO TAMBOPATA, MADRE DE DIOS, 2022"
Lugar	Jr. TUPAC AMARU,, DISTRITO DE TAMBOPATA, MADRE DE DIOS
Dist/Prov.	TAMBOPATA – TAMBOPATA
Solicitante	EDWARD JIMMY PANDIA YAÑEZ
Hecho por	ING. VICTOR HUGO CARAZAS MAYANGA
Fecha	20/06/2022

Datos de la Muestra

Calicata	P-1
Profundidad	1.50 m.
condicion	Alterada

Datos del Equipo Calibrado

Equipo	
PRENSA CBR	
Certificado de Calibración N° :	
MT-LF-054-2020 del 02/26/2020	

Datos y resultados de ensayo

C.B.R.	N° DE CAPAS : 5		
N° DE MOLDE	A	B	C
N° DE GOLPES	57	26	14
Volumen de Molde (cm³)	1985	1925	1919
Peso del Molde + Suelo Humedo (g)	11450	11001	10988
Peso del Molde (g)	7854	7945	7784
Peso del Suelo Humedo (g)	4364	4137	3912
N° Tarro	-	-	-
Peso Tarro + Suelo Humedo (g)	322.5	284.5	290.2
Peso Tarro + Suelo Seco (g)	280.4	269.3	279.2
Peso del Agua (g)	22.1	19.2	20.0
Peso de Tarro (g)	64.2	51.72	52.20
Peso del Suelo Seco (g)	239.2	215.3	224.2
Contenido de Humedad (g)	11.70	10.37	9.20
Densidad Humeda (g/cm³)	2.056	1.915	1.846
Densidad Seca (g/cm³)	1.789	1.636	1.586

PENETRACIÓN							
CAPACIDAD DE LA CELDA							
MOLDE		A		B		C	
PENETR. (pulg)	PENETR. (mm)	LECTURA DIAL	CARGA (kg)	LECTURA DIAL	CARGA (kg)	LECTURA DIAL	CARGA (kg)
0.000	0.000	0.0	0	0	0	0	0
0.025	0.63	86	86	50	50	38	38
0.05	1.27	100	100	69	69	45	45
0.075	1.9	157	157	130	130	90	90
0.1	2.54	180	180	145	145	120	120
0.125	3.81	298	298	185	185	150	150
0.2	5.08	280	280	190	190	198	198
0.3	7.62	390	390	290	290	286	286
0.4	10.16						
0.550	12.500						

ABSORCIÓN			
N° MOLDE	A	B	C
Peso del Suelo Humedo + Plato + Molde (g)	12820	12426	12035
Peso del Plato + Molde (g)	8426	8285	8070
Peso del Suelo Humedo Embebido (g)	4336	4141	3995
Peso del Suelo Humedo Sin Embeber (g)	4322	4136	3919
Peso del Agua Absorbida (g)	35	41	69
Peso del Suelo Seco (g)	3997	3784	3530
Absorcion de Agua (%)	0.9	1.3	1.9

EXPANSIÓN (%)						
FECHA	HORA	LECTURA DIAL	LECTURA DIAL	LECTURA DIAL	N° LECTURA	
		0.000"	0.000"	0.000"	1	
		0.001"	0.002"	0.002"	2	
		0.003"	0.002"	0.008"	3	
		0.004"	0.008"	0.010"	4	
		0.006"	0.010"	0.013"	5	
% EXPANSIÓN		0.13	0.19	0.25		

GEOTECNIA E INGENIEROS E.I.R.L.

VICTOR HUGO CARAZAS MAYANGA
INGENIERO CIVIL
C.I.P. 178442
AREA DE GEOTECNIA

LABORATORIO DE MECÁNICA DE SUELOS - CONCRETO Y MATERIALES - ESTUDIOS GEOTÉCNICOS (SUELOS Y ROCAS) - CONTROL DE CALIDAD DE OBRAS CIVILES
CONSULTORIA ESPECIALIZADA - PERFORACIÓN Y SONDAJE PARA ACUIFEROS Y CIMENTACIONES PROFUNDAS - HINCADO DE PILOTES - PROSPECCIÓN GEOFISICA
PUERTO MALDONADO JR CUSCO 134 - TAMBOPATA   CUSCO URB. BEZA REDONDA A-9 - CUSCO   982737067   083-574754   RUC : 20480331961

VALOR RELATIVO DE SOPORTE CBR (ASTM D1883-16, NTP 339.145)

Datos del poyecto

Proyecto	"ESTABILIZACIÓN DE LA SUBRASANTE DEL SUELO NATURAL CON ADICIÓN DE CAUCHO GRANULAR DE NEUMÁTICOS EN EL JR. TUPAC AMARU, DISTRITO TAMBOPATA, MADRE DE DIOS, 2022"
Lugar	Jr. TUPAC AMARU,, DISTRITO DE TAMBOPATA, MADRE DE DIOS
Dist./Prov.	TAMBOPATA -- TAMBOPATA
Solicitante	EDWARD JIMMY PANDIA YAÑEZ
Hecho por	ING. VICTOR HUGO CARAZAS MAYANGA
Fecha	20/06/2022

Datos de la Muestra

Calicata	P-1
Profundidad	1.50 m.
condicion	Alterada

Datos del Equipo Calibrado

Equipo	:
	PRENSA CBR
Certificado de Calibración N°	:
	MT-LF-054-2020 del 03/26/2020

Datos y resultados de ensayo

C.B.R.	N° DE CAPAS : 5		
N° DE MOLDE	A	B	C
N° DE GOLPES	56	29	15
Volumen de Molde (cm³)	2150	2150	2150
Peso del Molde + Suelo Humedo (g)	12590	12801	11666
Peso del Molde (g)	6589	5485	6070
Peso del Suelo Humedo (g)	4364	4137	3912
N° Tarro	-	-	-
Peso Tarro + Suelo Humedo (g)	302.5	289.5	300.2
Pesp Tarro + Suelo Seco (g)	280.4	269.3	279.2
Peso del Agua (g)	22.1	19.2	23.0
Peso de Tarro (g)	60.3	54.72	55.20
Peso del Suelo Seco (g)	239.2	215.3	224.2
Contenido de Humedad (g)	9.70	8.92	10.20
Densidad Humeda (g/cm³)	2.056	1.915	1.846
Densidad Seca (g/cm³)	1.789	1.636	1.786

PENETRACIÓN								
CAPACIDAD DE LA CELDA								
MOLDE		A		B		C		
PENETR. (pulg)	PENETR. (mm)	LECTURA DIAL	CARGA (kg)	LECTURA DIAL	CARGA (kg)	LECTURA DIAL	CARGA (kg)	
0.000	0.000	0.0	0	0	0	0	0	
0.025	0.63	88	88	60	60	42	42	
0.05	1.27	100	100	75	75	45	45	
0.075	1.9	157	157	130	130	100	100	
0.1	2.54	180	180	145	145	120	120	
0.125	3.81	298	298	185	185	145	145	
0.2	5.08	360	360	260	260	198	198	
0.3	7.62	525	525	390	390	250	250	
0.4	10.16							
0.500	12.500							

ABSORCIÓN			
N° MOLDE	A	B	C
Peso del Suelo Humedo + Plato + Molde (g)	11540	12426	13458
Peso del Plato + Molde (g)	8426	8285	8070
Peso del Suelo Humedo Embebido (g)	4336	4141	4995
Peso del Suelo Humedo Sin Embeber (g)	4322	4136	3919
Peso del Agua Absorbida (g)	36	39	69
Peso del Suelo Seco (g)	2997	3184	3530
Absorcion de Agua (%)	0.8	1.1	2.2

EXPANSIÓN (%)						
FECHA	HORA	LECTURA DIAL	LECTURA DIAL	LECTURA DIAL	N° LECTURA	
		0.000"	0.000"	0.000"	1	
		0.001"	0.002"	0.002"	2	
		0.003"	0.006"	0.014"	3	
		0.003"	0.008"	0.012"	4	
		0.008"	0.010"	0.015"	5	
% EXPANSIÓN		0.13	0.27	0.38		

GEOIN GEOTECNIA E INGENIEROS EIRL.
LABORATORIO DE MECANICA DE SUELOS - CONCRETO Y MATERIALES - ESTUDIOS GEOTECNICOS (SUELOS Y ROCAS) - CONTROL DE CALIDAD DE OBRAS CIVILES
CONSULTORIA ESPECIALIZADA - PERFORACION Y SONDAJE PARA ACUIFEROS Y ORIENTACIONES PROFUNDAS - HINCADO DE PILOTES - PROSPECCION GEOFISICA
(PUERTO MALDONADO JR CUSCO 134 - TAMBOPATA (CUSCO URB. MEZA REDONDA A-9 - CUSCO 982737087 063-574754 RUC : 20480031081

VALOR RELATIVO DE SOPORTE CBR (ASTM D1883-16, NTP 339.145)

Datos del proyecto

Proyecto	"ESTABILIZACIÓN DE LA SUBRASANTE DEL SUELO NATURAL CON ADICIÓN DE CAUCHO GRANULAR DE NEUMÁTICOS EN EL JR. TUPAC AMARU, DISTRITO TAMBOPATA, MADRE DE DIOS, 2022"
Lugar	Jr. TUPAC AMARU,, DISTRITO DE TAMBOPATA, MADRE DE DIOS
Dist/Prov.	TAMBOPATA — TAMBOPATA
Solicitante	EDWARD JIMMY PANDIA YAÑEZ
Hecho por	ING. VICTOR HUGO CARAZAS MAYANGA
Fecha	20/06/2022

Datos de la Muestra

Calicata	P-1
Profundidad	1.50 m.
condicion	Alterada

Datos del Equipo Calibrado

Equipo	PRENSA CBR
Certificado de Calibración N°	MT-LF-054-2020 del 02/26/2020

Datos y resultados de ensayo

C.B.R.	N° DE CAPAS : 5		
N° DE MOLDE	A	B	C
N° DE GOLPES	56	25	12
Volumen de Molde (cm³)	2050	2050	2050
Peso del Molde + Suelo Humedo (g)	12790	12401	11986
Peso del Molde (g)	8426	8285	8070
Peso del Suelo Humedo (g)	4364	4137	3912
N° Tarro	-	-	-
Peso Tarro + Suelo Humedo (g)	302.5	289.5	300.2
Peso Tarro + Suelo Seco (g)	280.4	269.3	279.2
Peso del Agua (g)	22.1	19.2	20.0
Peso de Tarro (g)	80.2	54.72	55.20
Peso del Suelo Seco (g)	239.2	215.3	224.2
Contenido de Humedad (g)	9.70	9.32	9.20
Densidad Humeda (g/cm³)	2.056	1.915	1.846
Densidad Seca (g/cm³)	1.889	1.736	1.686

PENETRACIÓN							
CAPACIDAD DE LA CELDA							
MOLDE		A		B		C	
PENETR. (pulg)	PENETR. (mm)	LECTURA DIAL	CARGA (kg)	LECTURA DIAL	CARGA (kg)	LECTURA DIAL	CARGA (kg)
0.000	0.000	0.0	0	0	0	0	0
0.025	0.63	86	86	50	50	38	38
0.05	1.27	100	100	75	75	45	45
0.075	1.9	157	157	130	130	100	100
0.1	2.54	180	180	145	145	120	120
0.125	3.81	298	298	185	185	111	111
0.2	5.08	360	360	260	260	198	198
0.3	7.62	500	500	355	355	286	286
0.4	10.16						
0.500	12.700						

ABSORCIÓN			
N° MOLDE	A	B	C
Peso del Suelo Humedo + Plato + Molde (g)	12820	12426	12025
Peso del Plato + Molde (g)	8426	8285	8070
Peso del Suelo Humedo Embebido (g)	4336	4141	3995
Peso del Suelo Humedo Sin Embeber (g)	4322	4136	3919
Peso del Agua Absorbida (g)	35	44	79
Peso del Suelo Seco (g)	3997	3784	3530
Absorcion de Agua (%)	0.8	1.1	2.2

EXPANSIÓN (%)						
FECHA	HORA	LECTURA DIAL	LECTURA DIAL	LECTURA DIAL	N° LECTURA	
		0.000"	0.000"	0.000"	1	
		0.001"	0.002"	0.002"	2	
		0.003"	0.006"	0.008"	3	
		0.004"	0.008"	0.012"	4	
		0.008"	0.010"	0.015"	5	
% EXPANSIÓN		0.16	0.20	0.30		

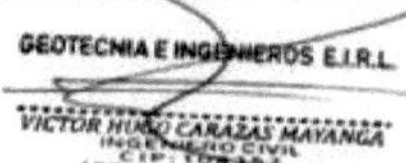

GEOIN GEOTECNIA E INGENIEROS EIRL.

LABORATORIO DE MECANICA DE SUELOS · CONCRETO Y MATERIALES · ESTUDIOS GEOTECNICOS (SUELOS Y ROCAS) · CONTROL DE CALIDAD DE OBRAS CIVILES
CONSULTORIA ESPECIALIZADA · PERFORACIÓN Y SONDAJE PARA ACUIFEROS Y CIMENTACIONES PROFUNDAS · HINCADO DE PILOTES · PROSPECCIÓN GEOFISICA
PUERTO MALDONADO JR. CUSCO 136 - TAMBOPATA · CUSCO URB. MEZA REDONDA A-9 - CUSCO · 982737067 · 082-574794 · RUC : 20480001061

VALOR RELATIVO DE SOPORTE CBR (STM D1883-16, NTP 339.145)

Datos del poyecto

Proyecto	:	"ESTABILIZACIÓN DE LA SUBRASANTE DEL SUELO NATURAL CON ADICIÓN DE CAUCHO GRANULAR DE NEUMÁTICOS EN EL JR. TUPAC AMARU, DISTRITO TAMBOPATA, MADRE DE DIOS, 2022"
Lugar	:	Jr. TUPAC AMARU, DISTRITO DE TAMBOPATA, MADRE DE DIOS
Dist/Prov.	:	TAMBOPATA – TAMBOPATA
Solicitante	:	EDWARD JIMMY PANOIA YAÑEZ
Hecho por	:	ING. VICTOR HUGO CARAZAS MAYANGA
Fecha	:	20/06/2022

Datos de la Muestra

Datos del Equipo Calibrado

Calicata	:	P-1	Equipo	:	PRENSA CBR
Profundida.	:	1.50 m.	Certificado de Calibración N°	:	MT-LF-054-2020 del 02/26/2020
condicion	:	Alterada			

Datos y resultados de ensayo

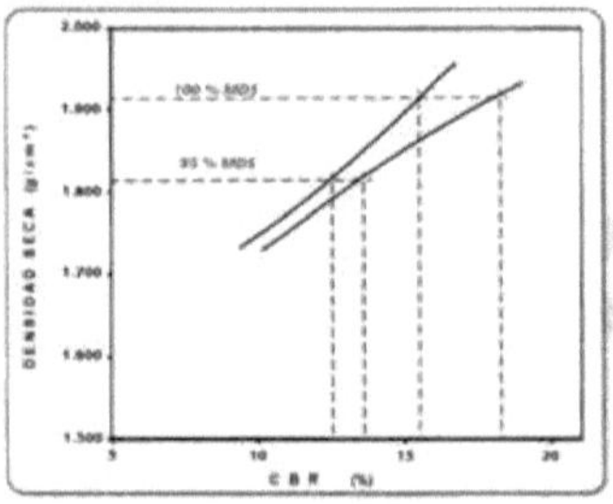

DATO DEL ENSAYO DE PROCTOR MODIFICADO
Optimo Contenido de Humedad (%) : 10.255
Maxima Densidad Seca g/cm^3 : 1.929

CALIFORNIA BEARING RATIO
CBR A 2.5 mm (0.1") de Penetración
CBR Al 100% de la Maxima Densidad Seca   10.1
CBR Al 95% de la Maxima Densidad Seca   18.9

CBR A 5 mm (0.2") de Penetración
CBR Al 100% de la Maxima Densidad Seca   14.1
CBR Al 95% de la Maxima Densidad Seca   11.2

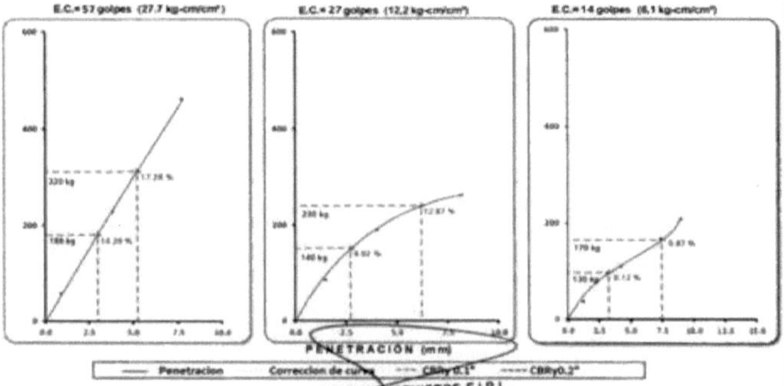

GEOTECNIA E INGENIEROS E.I.R.L.

VICTOR HUGO CARAZAS MAYANGA
INGENIERO CIVIL
CIP : 108191
AREA DE GEOTECNIA

VALOR RELATIVO DE SOPORTE CBR (STM D1883-16, NTP 339.145)

Datos del poyecto

Proyecto	:	"ESTABILIZACIÓN DE LA SUBRASANTE DEL SUELO NATURAL CON ADICIÓN DE CAUCHO GRANULAR DE NEUMÁTICOS EN EL JR. TUPAC AMARU, DISTRITO TAMBOPATA, MADRE DE DIOS, 2022"
Lugar	:	Jr. TUPAC AMARU, DISTRITO DE TAMBOPATA, MADRE DE DIOS
Dist/Prov.	:	TAMBOPATA – TAMBOPATA
Solicitante	:	EDWARD JIMMY PANOIA YAÑEZ
Hecho por	:	ING. VICTOR HUGO CARAZAS MAYANGA
Fecha	:	20/06/2022

Datos de la Muestra

			Datos del Equipo Calibrado	
			Equipo	:
Calicata	:	P-1	PRENSA CBR	
Profundida.	:	1.50 m.	Certificado de Calibración N° :	
condicion	:	Alterada	MT-LF-054-2020 del 02/26/2020	

Datos y resultados de ensayo

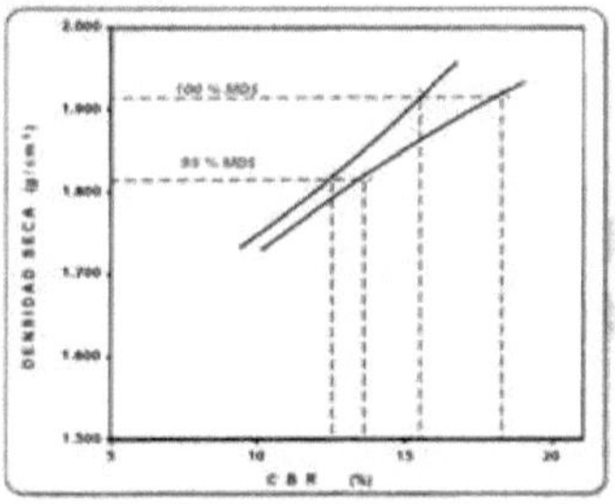

DATO DEL ENSAYO DE PROCTOR MODIFICADO

Optimo Contenido de Humedad (%)	:	10.255
Maxima Densidad Seca g/cm³	:	1.929

CALIFORNIA BEARING RATIO

CBR A 2.5 mm (0.1") de Penetración

CBR Al 100% de la Maxima Densidad Seca	10.1
CBR Al 95% de la Maxima Densidad Seca	18.9

CBR A 5 mm (0.2") de Penetración

CBR Al 100% de la Maxima Densidad Seca	14.1
CBR Al 95% de la Maxima Densidad Seca	11.2

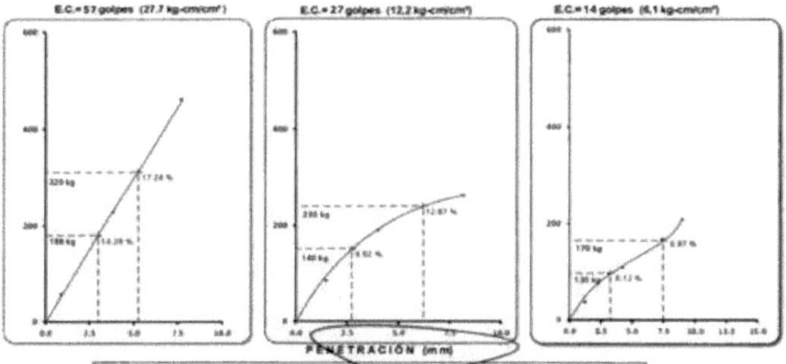

GEOTECNIA E INGENIEROS E.I.R.L.
VICTOR HUGO CARAZAS MAYANGA
INGENIERO CIVIL
CIP : 108192
AREA DE GEOTECNIA

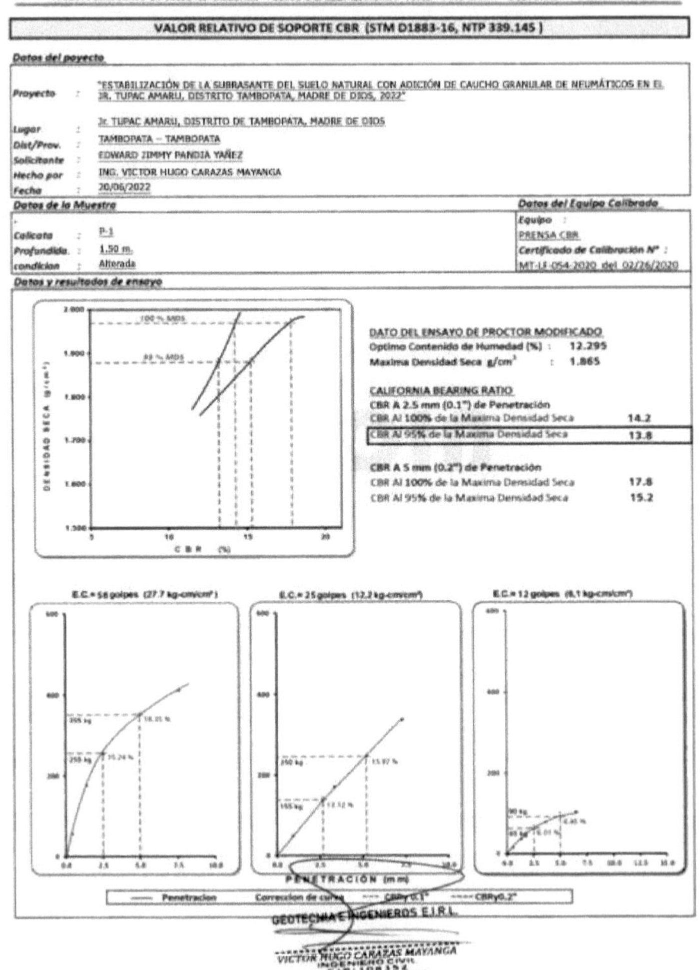

GEOIN GEOTECNIA E INGENIEROS EIRL.

VALOR RELATIVO DE SOPORTE CBR (STM D1883-16, NTP 339.145)

Datos del poyecto

Proyecto	:	"ESTABILIZACIÓN DE LA SUBRASANTE DEL SUELO NATURAL CON ADICIÓN DE CAUCHO GRANULAR DE NEUMÁTICOS EN EL JR. TUPAC AMARU, DISTRITO TAMBOPATA, MADRE DE DIOS, 2022"
Lugar	:	Jr. TUPAC AMARU, DISTRITO DE TAMBOPATA, MADRE DE DIOS
Dist/Prov.	:	TAMBOPATA – TAMBOPATA
Solicitante	:	EDWARD ZIMMY PANDIA YAÑEZ
Hecho por	:	ING. VICTOR HUGO CARAZAS MAYANGA
Fecha	:	20/06/2022

Datos de la Muestra

			Datos del Equipo Calibrado
Calicata	:	P-1	Equipo : PRENSA CBR
Profundida.	:	1.50 m.	Certificado de Calibración N° :
condición	:	Alterada	MT-LE-054-2020 del 02/26/2020

Datos y resultados de ensayo

DATO DEL ENSAYO DE PROCTOR MODIFICADO
Optimo Contenido de Humedad (%) : 12.295
Maxima Densidad Seca g/cm³ : 1.865

CALIFORNIA BEARING RATIO
CBR A 2.5 mm (0.1") de Penetración
CBR Al 100% de la Maxima Densidad Seca — 14.2
CBR Al 95% de la Maxima Densidad Seca — 13.8

CBR A 5 mm (0.2") de Penetración
CBR Al 100% de la Maxima Densidad Seca — 17.8
CBR Al 95% de la Maxima Densidad Seca — 15.2

GEOTECNIA E INGENIEROS E.I.R.L.
VICTOR HUGO CARAZAS MAYANGA
INGENIERO CIVIL
CIP 108 152
AREA DE GEOTECNIA

Annex N° 3: Photographic Panel

Material extracted from test pit N°1

Laboratory sieves - sieve particle sizing

Laboratory oven

Compaction test equipment

Field density equipment

Annex N° 4:

INSTRUMENT VALIDATION CERTIFICATE

JUAN FELIPE, RODRÍGUEZ PASCO, with DNI Nº 00371465 currently working as FULL-TIME

TEACHER at the UNIVERSIDAD ALAS PERUANAS FILIAL MADRE DE DIOS.

I hereby certify that I have reviewed the Bachelor's instruments for validation purposes.

Name of Instrument:

MOISTURE CONTENT
CONSISTENCY LIMITS
MODIFIED PROCTOR
RELATIVE SUPPORT VALUE CBR

After making the pertinent observations, I can formulate the following appreciations.

EVALUATION OF DATA SHEETS	DEFICIENT	ACCEPTABLE	GOOD	VERY GOOD	EXCELLENT
Clarity: It is formulated in appropriate language.			X		
Objectivity: It is expressed in observable behaviors.			X		
3.Actuality: Adequate to the theoretical approach addressed in the research.			X		
Organization: There is a logical organization among your items.			X		
5. Sufficiency: It includes the necessary aspects in quantity and quality.			X		

6.Intentionality: Adequate to assess the dimensions of the research topic.			X		
7.Consistency: Based on theoretical-scientific aspects of the research.			X		
Coherence: It has a relationship between the variables and indicators.			X		
9.Methodology: The strategy responds to the elaboration of the research.			X		

In conformity, I hereby sign this document in the city of Madre de Dios on the 30th day of the month of December, 2022.

Grade   Degree : Magister

DNI   : 00371465
Specialty   : Engineering
E-mail   : feropa57@hotmail.com

Mg. Ing. JUAN FELIPE RODRIGUEZ PASCO
DOCENTE
Código : 057142

yes
I want morebooks!

Buy your books fast and straightforward online - at one of world's fastest growing online book stores! Environmentally sound due to Print-on-Demand technologies.

Buy your books online at
www.morebooks.shop

Kaufen Sie Ihre Bücher schnell und unkompliziert online – auf einer der am schnellsten wachsenden Buchhandelsplattformen weltweit! Dank Print-On-Demand umwelt- und ressourcenschonend produziert.

Bücher schneller online kaufen
www.morebooks.shop

Printed by Books on Demand GmbH, Norderstedt / Germany